The Five Continuous Fundamental Charge-Magnetics Forces

Reconnecting Newton and Geometry into Chemistry and Particle Physics

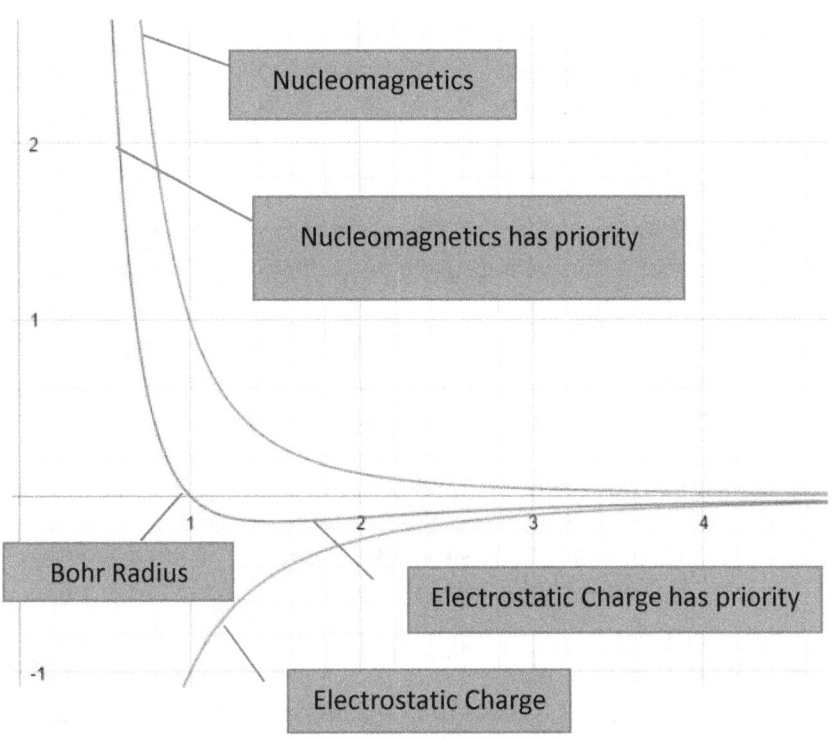

Resolving Strong Force, Weak Force, Bonding Force, Gravity, Mass and via the basics of Electromagnetism (Charge and Magnetics) as One Continuous Function

By Arno Vigen

© 2017/08 E Arno Vigen

Simple Words to Understand . . . Atoms and Chemistry

Why does a Nucleus Stay Together If Protons Repel?

- A Nucleus is Just . . . a Nucleomagnetics Ring

Why Don't Electrons Fall into the Opposite-Charged Nucleus?

- Electrons are Just . . . Frightened by Nucleus nucleomagnetics

Electron Shell Chemistry Is Just . . . Scrunched Cube Geometry

- Why are electron shells in sets of 2, then 8, then 8 and such? Can we improve Pauli-aufbau?

Scrunched Cube Periodic Chart of Elements

- What are the properties and groupings of elements using the Arno Vigen Scrunched Cube model

Scrunched Cube Chemical Bonding

- Why does Nitrogen and Oxygen have Different Bonding Angles? What drives bonding?

What Makes a Molecule Solid, Liquid, or Gas?

- And Why is the Gas of Every Element the Same Volume (a mole)?

Simple Words to Understand . . . Gravity, Electromagnetism, and Other Forces

Gravity is Just . . . That Electrons are a Little Closer

- Explaining Gravity from the basics of Electromagnetism and Explaining Why Observed Mass Changes

Does Time and Space Really Warp?

- Replacing Electron-Shell Radius for Time-Space Factors in formulas such as the General Theory of Relativity

How are Electricity and Magnetism Linked?

- Exploring the Fundamental Linkage of Charge and Magnetism

Fixing Einstein's $E = mc^2$

- Given mass $m = \dfrac{M(z,n)}{\frac{8}{3}\pi (R_{ES})^3}$ defined by AVSC nucleomagnetics, then what is mc^2

Beyond the Quantum Era

- Restoring Newton, and Moving Beyond Quantum Mechanics

Simple Words to Understand . . . Personality

Visual Astrology: Fun, Support, Security, and Growth

- Astrology 'signs' archetypes are based upon powerful traits to understand people

Visual Astrology Relationships

- What happens when 'sign' personalities interact

Visual Astrology and Jung

- Astrology 'signs' archetypes actually predict all the Jungian 4 archetypes

Dominant Personality Traits

- Dominant Personality Traits Follow from Four Dimensions, Six Steps and so 24 Subcategories

Simple Words to Understand . . . Communications

Decision Matrix® Writing

- Persuasion is based making arguments at the correct strength in a certain order.

GATESOUP® Writing

- **G**oal, **A**udience, **T**heme, Enough **E**lements, **S**upport and the rest

Kedarf® Grammar and Composition Explained

- Defining the Parts of Speech, Paragraph Structure and More in Usable Terms

Table of Contents

Making All Levels of Physics, Chemistry Follow Newton's Laws... 11

    The Challenge 11

    Criteria for Successful Postulates 14

    Additional Major Challenge – Only Charge and Magnetics and Known Particles; Only Protons, Neutrons and Electrons 17

    Methods to Resolve Any Discontinuities 19

        Defection, bumping off each other occurs before two particles actually occupy same physical location. Particles have physical dimensions 20

        Forces Balance – Action and Reaction 21

    Visual Discovery for Understanding the Challenges 23

    Understanding that Particles Can Give Double Forces – Charge and Magnetism – Creating Reflection Balancing Points and Limits 27

Fundamental Basics of Charge-Magnetics 38

    Both Charge Force and Magnetic Force Balance: Opposites Attract, Like-Types Repel 42

    Settling on a Presentation of Charge, Magnetics and the Net Charge-Magnetics Forces 45

    Correct graph has those charge and magnetism functions as Newtonian opposites 45

    Postulate #1: Charge and Magnetism are the Newton 3rd Law opposites 46

    Corollary #1A: Charge forces and its activity (movement) exists with corresponding Magnetism effect. One reacts to the other. Always show charge and magnetics with opposite signs in calculations 47

Charge and Magnetism in Combination – the Basic 'Net' Force .. 49

    The Fundamental Force: Netting the Force of Charge and the Force of Magnetics.. 54

    The Net Charge-Magnetics Force Does Have Distance Difference Profile 55

    At Each End, Charge and Magnetism Mask That the Function is the Combination 56

    Postulate #2 – The Balancing of Charge and Magnetism Is Exactly What Creates Electron Shells; Net Charge-Magnetics Is the Weak Force. 60

    Postulate #3 - That Point of Balance Is the Integral of Magnetic Nucleus Particles (ZP) Strength Force At the Average Bohr Radius Distance 61

Fundamental Charge-Magnetics Force: The Net 64

    Repulsive at Close Distances and Attractive at Large Distances 64

    Fundamental Charge-Magnetic Force: The Net Viewed for Opposite Force – Charge versus Magnetics 65

Graph of the Charge-Magnetics of a Proton................................ 68

Graph of the Charge-Magnetics of an Electron 69

Graph of the Charge-Magnetics of a Neutron Leads to Strong Force that Holds Every Nucleus Together................................... 70

    Magnetic Chains Proton-Neutron-Proton Strong Force 72

    Magnetics Stay Strong in Chains... 72

    Postulate #4 – A Nucleus Physical Structure, Proton-Neutron-Proton, in a Magnetic Chain or Chain-Ring or Combination Has Magnetics That Stays Strong, and Enough Distance Between Protons So That Proton-Proton Charge Repulsion Has Decreased. To Remain Stable, a Nucleus Structure Must Have At Least One Neutron As Physical Separation Between All Protons. The Entire Magnetic Structure Is Stable As Magnetics

Is Stronger Than Proton Charge Repulsion At Those Distances. 72

Understanding the Magic Balancing Point 76

    How to Make Radius of Electron Shell into a Constant 77

    Postulate #5: The Volume of the Electron Shell has a Direct Correlation to the Number of Magnetic Nucleus Particles. (With very minor adjustments for a) internal cancellations for the nucleus chain-ring shape, structure called 'mass deficit' or b) for any the magnetics of electrons not contributing because they sit between two nuclei in bonding 'bonding mass reduction') 78

    Comparison to Bohr 80

    Magnetic Field Is Stronger Inside the Electron Shell, Charge Stronger Outside the Electron Shell 84

    Fundamental Question: Why Doesn't Proton Charge Repulsion Make Nucleus Fall Apart? 84

Measuring Charge-Force versus Magnetic-Force at Distance in Nucleus 88

Defining Particle Forces Versus Combo-Forces 93

    Corollary #5A: Strong Interaction Force and a Gluon is Really a Combo Force of the Combination Closer Magnetics-Only of a Neutron and the Net Charge-Magnetics of a Proton Separated by the Neutron, but Connected by Magnetics, so a Stable Structure 98

Gravity As the Charge-Magnetics Attraction Force of Electrons Being a Little Closer than That Repulsion from Protons. 99

    1/Distance-Squared ($1/d2$) goes forever and decreases exponentially 100

    Electron-shells is Another Part of the Equation 102

    1/distance-squared makes closer items more powerful 103

    'Big' Charge versus Magnetics versus 'Tiny' Gravity 104

Postulate #7: Net of Charge of Electrons in the Electron Shell as Integral Over Time Versus the Charge of Protons in Nucleus is the Gravity Fundamental Force 105

Calculation of Gravity from Charge versus Newton 6.674×10^{-11} $m3kg(s2)$ 108

Gravity Using the Graphics of Charge-Magnetics 111

Electron Shell Balancing Weak Force Important to Gravity.... 112

Postulate #7: Electrons, while attracted to charge, are repelled by Magnetics, Both North and South 113

Corollary #7A: Electron Repulsion to the Magnetic Field of Nucleus Particles Is the Weak Force 115

Question. What is the strength of that Repulsion? 116

Postulate #2 Revisited: The charge force and magnetic force balance each other as Newtonian opposites at the average volume or distance of the Electron Shell RES. 116

Most Fundamental Constants Will Consolidate with a Factor Directly from the Average Volume or Distance of the Electron Shell RES. 117

Nucleus Binding 'Strong Interaction' Force Calculation .... **Error! Bookmark not defined.**

Postulate #5 - Revisited: The Strong Forces is Magnetics Extended in Proton-Neutron-Proton Chains So Separate of Neutron a) Keeps Magnetics Strong, but b) Allows Enough Reduction of Proton Charge Repulsion .... **Error! Bookmark not defined.**

A Stable Nucleus Exists Because Magnetics Maintain Strength in Chains, Overcoming Proton-Proton Charge Repulsion in a Nucleus by at Least one Neutron Separating Every Proton. **Error! Bookmark not defined.**

Fundamental Question: This Nucleus Distance Is a Significant Link Between Charge and Magnetics Direct Interaction and Creation 119

    Challenge: The Balance in Strength is NOT at the Edge of the Particle 119

    Fact: At the Particle Edge, Magnetics are Stronger 119

    Fact: At the Particle Edge, Magnetics are Stronger By the Exact Factor of the Charge/Magnetics 119

Question: What is Stronger Over all Distances? Charge or Magnetics for One Particle .... **Error! Bookmark not defined.**

Molecular Bonding Force 122

    Graph Needs Electrons Offset to Avoid Discontinuity (the Deflection of Directly Hitting Electrons in the Shell) 125

Connection of Charge-Magnetics to Mass 133

Postulate #8: The Fundamental Charge-Magnetics Function is a double-function operating from and delivering results to both charge and magnetism as two output values in coordination. .. 134

Newton's Gravity Formula Revised 136

Einstein's Energy Formula Revised 140

    Why does Einstein say that space and time warp? 141

    Fundamental Question: Why does 'mass' change at speeds near the speed of life? 142

    Fundamental Question: What Explains why Mass Changes in Bonding? 143

        What changes is RES, not (ZN) (the number of nucleus particles). 143

    Bonded Electrons do not contribute to the external Magnetics (Atom-gravity) 'mass' observed 145

    Decrease offset in that Electron still contributes to push out other electrons 147

- Decrease offset by increases in distance to electron in other atom in the bond 148
  - Must also realize location is not fixed, but a quantum path .. 150
- Subatomic Particle Forces 151
- What is a Neutron? 155
  - Fundamental Challenge: All Work in Subatomic Particles Done in Charge (eV), but Not Enough Done on Magnetics? 159
- Atomic Mass Used in Many Force Calculations 160
- What is the magnetic field of a nucleus particle/nucleon (proton, neutron)? 161
- What is a Neutron? 163
- Question: Why is it important the Magnetic Force Overwhelms Charge Force at the Sub Nucleon Distance Range, but Multiple Charges are Not Observed? 164
- $E = mc2$ and the Fundamental Ratio of Electromagnetism 166
- Education Opportunities Based upon Postulates 168
  - Resolving and Restating Mass Changes Everything 168
  - Reintroducing Geometry Replacing Statistical Solutions 169
- Conclusions and Opportunities 171
- Arno Vigen Science Postulates: 173
- Endnotes 181

Making All Levels of Physics, Chemistry Follow Newton's Laws

The Challenge

There has been much discussion, even new standards, in the scientific community, for a century, that for certain situations both common sense and the Newton Laws of Motion do not apply. Essentially, to resolve this, to solve those same challenges in a way that maintains Newton's Law, I approach the challenge with a fresh viewpoint. That view, with its own postulates, creates the fundamentals needed to clarify and resolve all fundamental forces into Newtonian basics.

In this section, I will set the framework that defines success for the set of postulates that follow. There is a high bar for what I am trying to accomplish; it goes beyond connecting the fundamental forces into one formula. The challenge is connecting them while also following all of Newton's classic laws.

The goal is to replace, with one base formula, the partial solutions that I see in textbooks with 'elementary particles' and their forces that I see shown as 'facts' in Wikipedia:

Force	Particle Causing	Range	Lifetime in Experiments
Weak Force	W and Z boson	~10^{-12}m only	3×10^{-25} seconds
Strong Force	Gluon	~10^{-15}m only	Nanoseconds
Gravity	Higgs Boson or graviton	>10^{-10}m	No successful experiments

Pardon! A life in nanoseconds above seems a huge disconnect to common sense. Even God cannot hold Carbon C-14 atoms together for millions of years with particles and forces that a century of experiments show can exist only for a fraction of second. The particles and forces must last for very long period, and the forces must work consistently over all distances.

Common sense tells me that at least some of these 'elementary particles' are transitory-particles (for a short period of time) and the observed interaction are combo-forces of more fundamental particles that get used for specific scientific calculations. That these special particles work only when needed for a small range of distances is a red flag. The real answer is the corrected combination of the base electrostatic charge and magnetics from known particles, like proton, neutrons, and electrons. A stable combination from those will last for long periods and still provide the specific attributes ('spin', 'color' and such).

I compare these observed, transitory particles and forces to a 'kiss'. You see a kiss is real, but brief. We crave kisses; we want more of them; we swoon when we discover them, and definitely want to learn how they work, and how to make them work better. However, the kiss is not the human, and you need real humans to create a kiss. A kiss only occurs when two humans interact. It is nothing without those big entities. A kiss can last a few seconds, but that human can last a century. In these postulates, we will focus on a similar big picture – charge and magnetics – to understand those two, and to find the other continuous fundamental forces. Those other fundamental forces are really connected to the big picture of the charge and magnetics of protons, electrons, and neutrons.

The smaller stuff, the kisses, will work itself out when the big picture is correct.

The main goal is to consolidate the fundamental forces into a combination of charge and magnetics. It is my attempt to get away from a dozen separate 'funny named' particles, and separate 'named' forces, and to show that all fundamental forces work from the same constants.

Criteria for Successful Postulates

Here are the measures of what I think would define a successful set of postulates and address the challenges to build that integrated system:

1. The five fundamental charge-magnetics forces are the parts of one continuous[i] function – based upon charge and magnetics. There are no magic particles that make a force go in a different direction or a force that starts and stops just for the time when needed to solve a particular situation. Similarly, there cannot be places where a force starts out of nowhere without a knowable particle creating it.

I am concentrating on the formal math meaning of 'continuous'. It is a high-hurdle technical standard.

Of course, each situation or observed other force has obvious differences of the number/type of particles, their properties, distances, and such. The situations of the five continuous fundamental forces are different, so the use of the formula will follow the situation. A force from two different particles can connect and relate uniquely to create different observations versus a force when three interact.

2. From Newton's 3rd Law of Motion, "every action has an equal and opposite reaction." Each force has a corresponding counterforce. There must be balance in nature. Yin and yang. The successful model of the fundamental charge-magnetics force should show

where the opposites exist, even if currently not directly part of the derived, observed 'force'.

In a vector diagrams, it means that there is always a reaction in the opposite direction. They can be different types, and even fat and wide versus long and skinny, but the total strength nets to zero. That is the orange and purple are actually the same total energy in the below example.

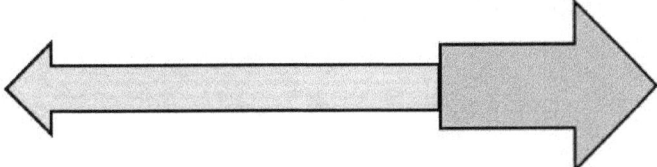

They can also be their funny multiples. Two might go off, at different angles, so a third release creates the total Newtonian net-balance combination.

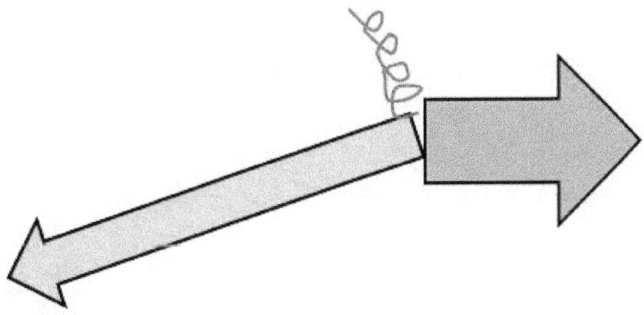

In fact, there is a famous picture from 1970 that shows exactly that type of reaction. Since the forces go at an angle, an extra element occurs so the entire interaction follow Newton's 3rd Law of Motion.

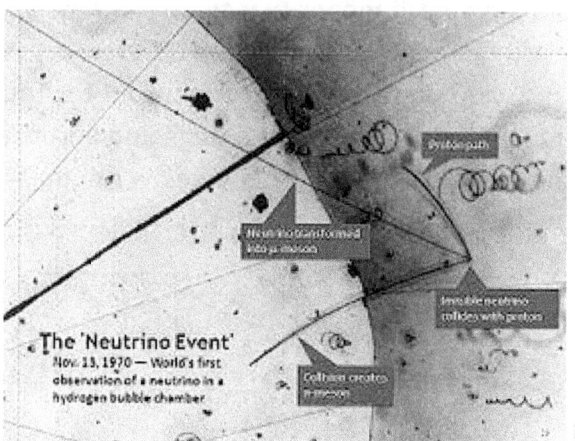

3. Those forces must work over continuous space and continuous time. Those five fundamental forces last centuries continuously, so the solution must have that same continuity. The solution cannot introduce other dimensions, attributes and such.

Those known dimensions are:

- Space (length, width, height or their spherical equals – radius, inclination/longitude, latitude/azimuth)

- Time (as a continuous fundamental)

- Energy (charge, magnetics, and their continuous strength fields)

The challenge is to solve all the basic, stable interactions within those considerable constraints.

Additional Major Challenge – Only Electrostatic Charge and Magnetics and Known Particles; Only Protons, Neutrons and Electrons

So, this challenge is enormously worse. All the other forces can only have a source from charge and magnetism, and at the critical point, charge and magnetism must act as the offsetting forces to resolve Newton's 3rd Law.

4. Solutions uses only two forces - charge and magnetics – and uses only the long-stable particles of protons, neutrons, and electrons.[iii]

Further, those particles cannot change their strength fields over different distances. The forces that solve a 'strong interaction' binding force of the nucleus binding distance must calculate must use the same constants to calculate the electron shell 'weak force' which works at a >1,000x different distance.

A solution set that has all four criteria above I consider a major success. If any one of the postulates replaces current beliefs about even one of the special elementary particles and forces, replacing a current textbook with a better term that contains a clear link to charge and magnetics, we will have move the pendulum back toward Newton, my hero. I believe that at least two of the follow postulates powerfully resolve the current model into a cleaner, simpler understanding.

Of course, it will take years to test every potential challenge, but first let's understand the postulates.

Methods to Resolve Any Discontinuities

Each of those goals have challenges. The challenges, of course, have solutions in physics and math. For most people, those math-type formulas are a challenge. I will follow a different path. I love geometry, and that becomes a picture, so hopefully, that presentation can connect with a broad set of people. Please bear with me as my approach comes from geometry, so the presentations will tend towards that type of analysis. My visual, geometry way of thinking, the correct solutions will have clear presentation in force graphs and vector diagrams. I like working visually, and let me explain why.

Discontinuities Easy To View

Let's discuss the first issue, that the solution force and its formula must be a continuous function. That is the graph of that force function must always change without jumps.

If a jump occurs, then typically each discontinuity usually flows from a function with $\frac{1}{x^n}$ such that if x is ever zero, then the result of $\frac{1}{0}$ is infinite. That is a break (yellow circle) in a function like the below.

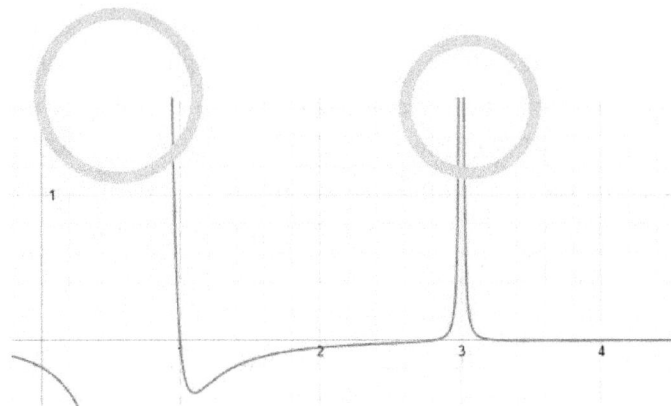

Defection, bumping off each other occurs before two particles actually occupy same physical location. Particles have physical dimensions

These discontinuities problems get resolved by:

- The distance is less than the physical dimensions of the particle. The particles bump before that sort of off the charts occurs. The results will become hugely repulsive before you could get to a zero distance.

    We see this result resolved in the real world, it is called nuclear explosions or nuclear decay. If the proton gest close then the force goes off the charts, but that occurs before it gets infinity. More commonly, each situation has a path around that zero where the two particles 'bounce' or deflect off each other. The interesting part is finding that calculation. Each of these can get resolved.

- The diagram only uses one (1) dimension, the distance of the observation, but the real situation actually has another direction where the force deflects the particle. Remember that 1970 picture. In this way, we need that extra dimension in the formula. For the stable forces in focus here, those forces do not have these exceptions. Therefore, for the basic forces discussed here, the graph of those forces cannot have $\frac{1}{0}$ because it must solve atoms that remain stable for centuries without 'bouncing' or deflecting.

Forces Balance – Action and Reaction

For the second challenge, Newton's 3rd Law of Motion, offsetting forces for every force, the postulates identify a series of balancing points exist with each force where that push and pull of charge and magnetism are obvious. When that action and reaction pair is not complete, those combinations will have an extra result, usually in magnetism, applying in a different direction. Lots of these are emissions, like photons, that release that unbalance in true Newtonian fashion. Again, I will resolve the big picture here, and the small balancing reactions, like photons, will work themselves out in a follow-on text.

Like the 1970 drawing, the solution, the full reaction, probably is another element that non-Newtonian solutions are not including. Often these extra reactions are immaterial, so no one finds them for most problems (charge and distance, ignoring magnetics being the main example); this is just expediency. Other times, these extra reactions are of a different type, so scientists do not observe them (nucleomagnetics being the main example).

In these postulates, there is charge and magnetics. Yet, 99% of scientific measurements focus on charge. The reason is simple. There is almost no method today to view the magnetics of a single proton. Which way is north? Where is another particle, which we would also need to know the north-south, to understand what magnetics is doing within this interaction? Magnetic is more difficult to find, understanding, and manipulate.

As you go through this postulate, critical points of balance have magnetics that when added to the postulate create a Newtonian system.

Visual Discovery for Understanding the Challenges

I am visual, so this exploration will take a very different path from others. Each step gets to a visual picture, a graph of the force versus the distance or a vector diagram. The purpose is that the graph identifies a number of very important locations.

In visual presentations, even of the same situation from different combinations, an explorer sees a bigger picture:

- Locations Where Forces Balances – By Intersections When Two Forces Graphed By the Same Measures

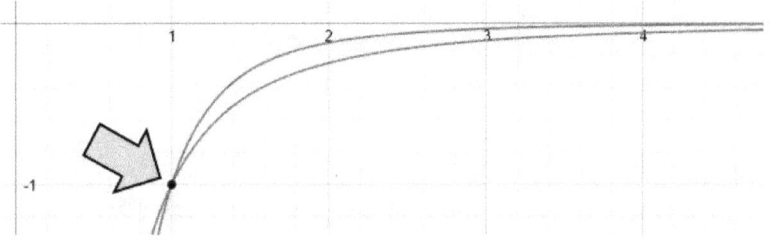

Or as A Zero Intersection for a Combined Force Graph

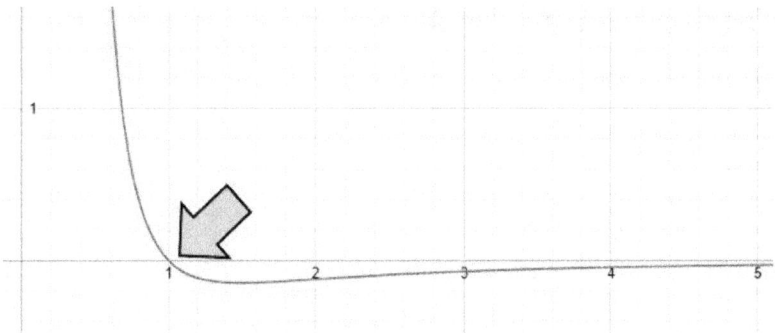

- Locations Where the Interaction is Maximum or Minimum (by the location where a compound formula's slope changes Positive to Negative or vice-a-versa)

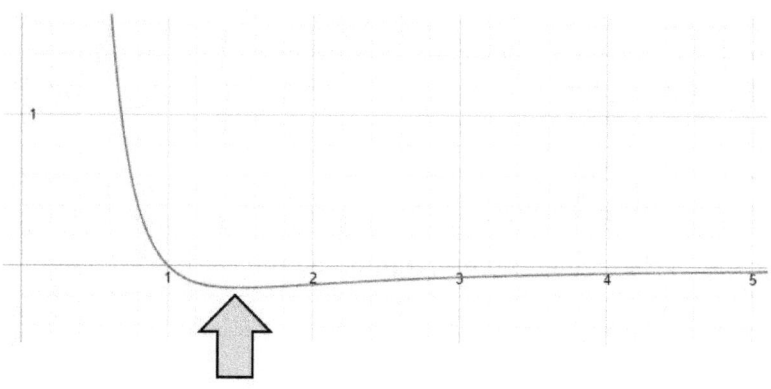

- Consistency of the Interaction in that the Force is a Continuous Line. No breaks, and no dramatic changes of slope (the change in the strength)

The below graph fits that solution. It is 'math' continuous.

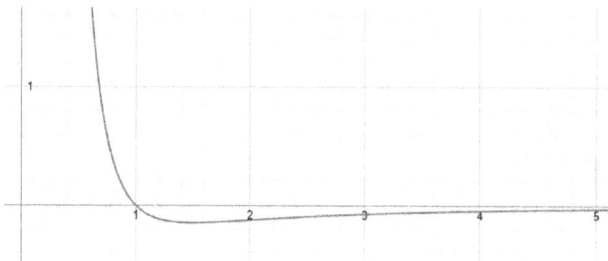

The most basic challenge is that in poor formulas we would see forces that jump, veer off, and even reverse and calculate forces based upon locations that do not have a particle. It should get a) based upon the balance of charge and magnetism, and b) based upon the relative location and orientation of the particles themselves.

A poor formula might also have a force that starts from nowhere without a particle at that location. For example, in the above graph, I find textbook presentations that show only the negative portion. The conclusion is that the force magically starts at Point 1, as if distances that repulse are not important. Of course, the full picture is that the other portion is not expressed in charge, there is a magnetic graph just as important. The challenge is to create the graph expressed in magnetics or things like neutrino (or photon) release in the reaction direction as per Newton's 3rd.

Most complex, often the initial assumption is that where the slope reverses that there is a particle at that location. That is certainly true of the discontinuities used in the first graph.

This graph shows the force of two particles on a 3rd in a straight line, not 3D. The one, opposite charged particle sits at zero and the like-charged particle sits at 3. You can see the graph goes off the charge at 0 and 3.

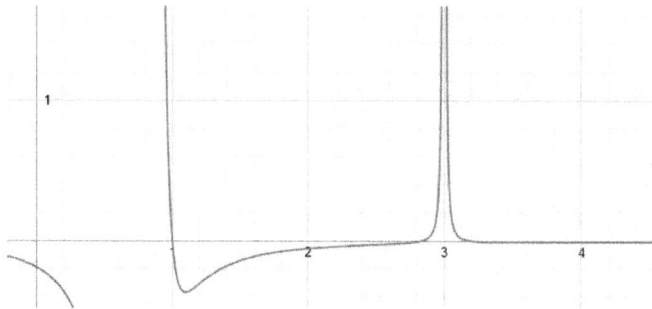

However, we math types will find that often there are reflections from where these force balance or maximize. These have location that scream to math wiz's these type of reflection points. Look at the below max which is located 1.5 past the balance point. There are math numbers like $\sqrt{2}\ or\ \sqrt{5}$. There are other like the golden ratio, 1.618xxx, but I will not spend much time of the intriguing mathematics oddities. The real purpose is finding the underlying fundamentals that create balance and maximum at certain distances. Again, the big picture first.

The special places that work 'at a distance' is the amazing part of the fundamental forces. Below is the most basic one. It is a graph of two Newtonian action-reaction forces that creates a reflection change of direction as a distance that we can calculate.

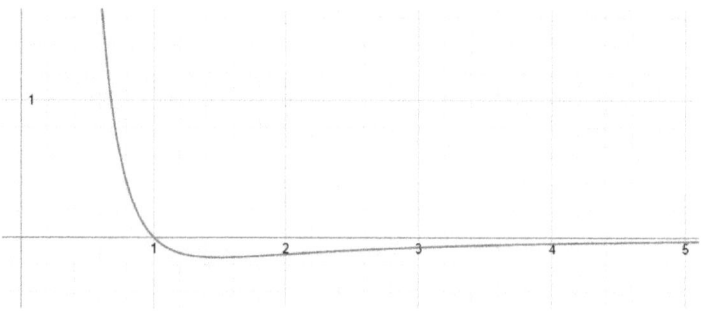

Understanding that Particles Can Give Double Forces – Charge and Magnetism – Creating Reflection Balancing Points and Limits

The graph before shows the reflection of two intersecting forces – charge (electrostatic) and nucleomagnetics. You will find this occurring often. Something exciting occurs when those two different forces intersect or balance.

There are basic standards driving these that were discovered and explained by Planck, Bohr, and Einstein. It is the main reason that I use the normalized 1,2,3 scale for most of my presentations. The scale is small, but the reflections come at understandable intervals.

I have tried to separate a) the common-sense part of how the situation combine forces from b) the actual distances based upon the basic standards. The visual is to get through the common sense first. Once done, then are the deep formulas comprehendible.

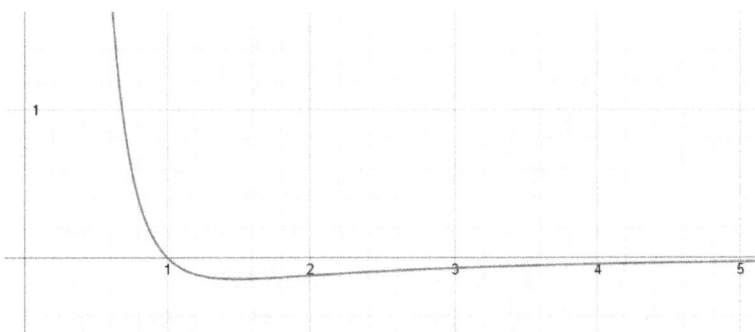

Understand that the above is really the showing the difference in the two below functions.

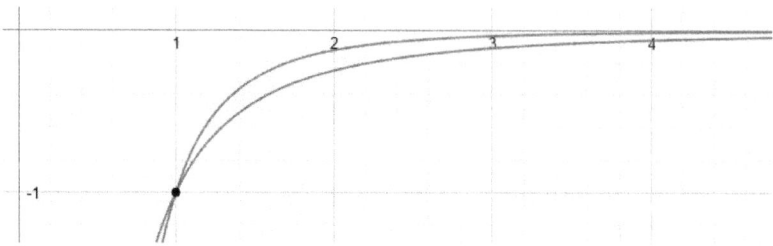

That means that we need to explore and discover where additional particles operate and from which forces these reflections derive. We need to explore the specialty locations, like where force = 0; where slope = 0; and where the change (Δ) in slope = 0, but keep digging until we have root cause for each part at each distance.

It is a herculean effort because, of the millions of calculations, it would only one discontinuity to create a break in those fundamentals.[iv] It would only take the requirement for one extra particle or one extra force or one extra attribute to break this model. I cannot cover every example here, but the postulates following do connect the fundamental forces to a continuous combination of charge-magnetics – charge and magnetics – from the basic particles.

Finally, if this postulate passed those tests, then the exploration 'lifts some of the curtain' on an idea that the charge and magnetics are the connected opposites in Newtonian fashion. Further, it gives directions to explore about why and how charge and magnetism have this symbiotic relationship. That is not one of these postulates, but a solution here would reframe more work into the charge-magnetics relationship in the ways that these combinations relate and resolve the fundamental forces.

Common sense tells me this theoretical direction. It calls to me that the universe must get simpler and more fundamental at the smallest layers. That is the best solutions have less particles that current thinking.

With some more efforts, probably in a later book, many of the current sub particles will become simple combinations.

If you are ready for adventure, let's explore. Big hugs, and let's get going.

Fundamental Basics of '-Magnetic' Fields

In AVSC, I take a different approach to the electrostatic and magnetic, or more specifically nucleomagnetics, fields. Prior scientists have called it 'electromagnetics', with the electrostatic charge well defined. However, it is magnetics that lacks a core function.

What causes a magnetic field?

How do we calculate it from base particles?

What is that relationship between nucleomagnetics and electrostatic charge that make them so intertwined, yet different?

I present a new postulate on magnetics that "lifts that veil"[v] on that fundamental force – magnetics. It may be "balancing on the dizzying path between genius and madness."[vi]

First, in the previous section, we discussed two Newtonian, opposite forces of different types. That chart is electrostatic charge force, which changes by 1/distance-squared, and nucleomagnetics force, which changes by 1/distance-cubed.

Postulate #1 – Electrostatic charge force changes by 1/distance-squared, versus 'nucleomagnetics' force changes by 1/distance-cubed with an axis and additional strength factor based upon the inclination (longitude) angle.

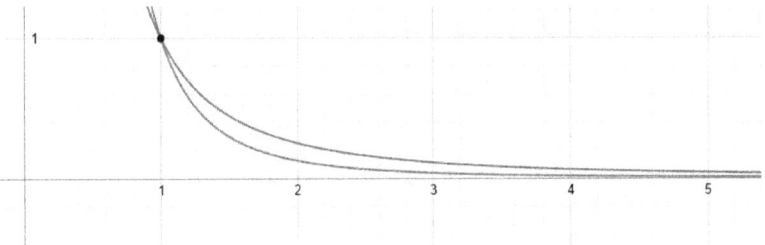

The critical understanding of this dynamic is decrease. At long distances, 1/distance-cubed becomes smaller faster. Therefore, at distances, nucleomagnetics is exponentially smaller, and thereby ignored. This why no scientist focused or found it before now.

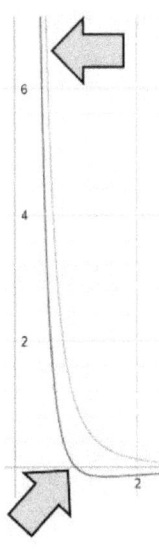

However, the fundamental misunderstanding of 1/distance-cubed, that is, nucleomagnetics, is that at very short distances (subatomic distances), magnetics is <u>actually larger</u>. That is the nature of 1/anything. It goes to infinity at a close distance.

In fact, at close (less than nucleus particle size) distances, the nucleomagnetics force is so strong, that electrostatic force become immaterial.

Therefore, at subatomic ranges, this nucleomagnetics, and not electrostatic charge, it a) very powerful – by the 1/cubed power, and b) more important than electrostatic charge – even if electrostatic charge is also 1/squared in strength.

All current calculations of subatomic forces and particles miss this when they use 'eV', electron volts, as the measurement. Volts is electrostatic charge, and when you measure it in the strongest power, nucleomagnetics, a scientist get inconsistencies. It is measuring the less important factor.

> By the way, that means when scientist measure subatomic particles in eV (electrostatic charge), they miss the bigger picture. Further, the scientists then wonder why the subatomic particle forces have strange interactions versus their 1/distance-squared calculations. That is because they are using the wrong formula, exponent, and orientation analysis method.
>
> That is like measuring the temperature of a hurricane versus the wind speed.
>
> "Oh, it is 70 degrees today."
>
> "I don't care, the wind is blowing 200 kilometers per hour."

Fundamental #2: Magnetics has an axis and strength based upon the inclination (longitude) angle, with the same strength at every latitude (azimuth).

All the '-magnetics' forces have an axis. We all know the magnetic fields for shape.

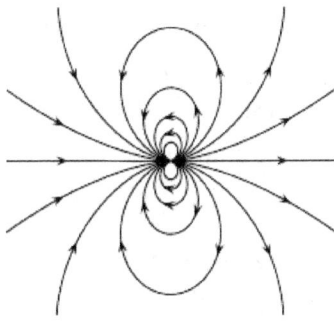

New Fundamental Basics of Magnetics Fields

Movement magnetics, that is, motomagnetics, is fundamentally different than the particle forces.

In math and science, we have three different attributes in most things.

- Location
- Speed
- Acceleration

If I am in a vehicle, I have a location. At Main and First Street.

If I am in a vehicle, I have a speed. It is moving at 40 kilometers per hour (for us backwards people in the USA, 25 miles per hour).

If I am in a vehicle, acceleration can the weirdest. The vehicle may be decelerating (slowing down), but I think I am getting pushed forward. In that way, the third attribution, acceleration, can express pushing, when the actual action is pulling. It can have a weird north-south dynamic as I will show over the next few pages.

While not a perfect analogy, this the type of structure that describes the AVSC Atomic Model for the fundamental 'electromagnetic' forces. By that word, you think there are two forces, but really there are three expressions of the the fundament force:

- Electrostatic Charge

- Nucleomagnetics

- Motomagnetics (traditional, big-world 'magnetics')

Now, the analogy is not in a line, like location, speed, and acceleration. It is for a field of force in 3D space – in which the strength has that crazy inclination angle change. Yet, the fundamental 'electromagnetic-3' acts by the three-layer model. This goes to the basic nature of fields:

- Electrostatic Charge is the force increase (think 'up') of the charge versus the neutral state (Coulomb)

- Nucleomagnetics is the constant 'pull' (think 'down) in the universe to get the 'overall field' back to the neutral state (AVSC)

- Motomagnetics (traditional, big-world 'magnetics') is the change in the field which is a different pull to get the local area at the same 'field strength' even if it is not zero. (Gauss)

That means that we need to think about the 'electromagnetics' field as always having two dynamics. The first dynamic of the charge (electrostatic) versus Newtonian opposite (nucleomagnetics) force to return the overall field to neutral. The second dynamic is a time-sensitive field that wants to bring keep the force at whatever state it just was.

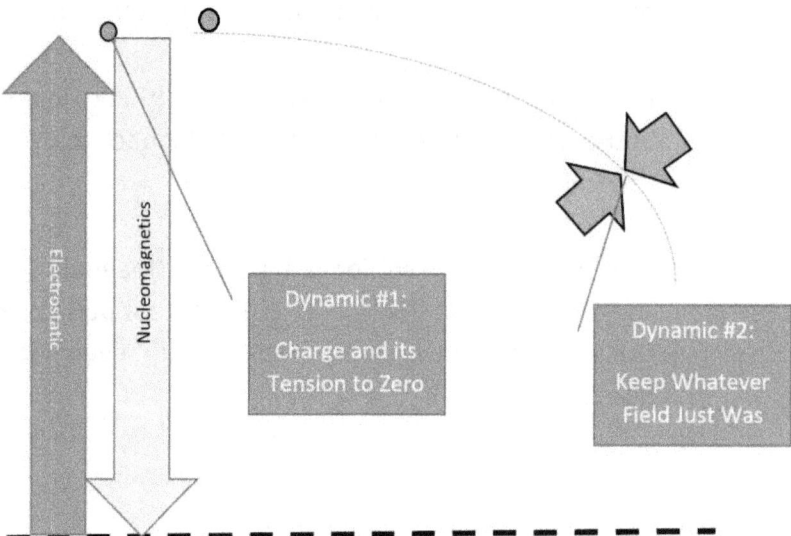

The comparison of magnetics to Newton's First Law is very interesting. Gauss has shown that the total charge of a region of space must have the same total magnetic field as that group over space and time. Yet, the magnetics and particle can move.

Gauss worked to resolve, document and calculate the second dynamic.

This is like speed and acceleration. You can go one speed, but that can increase or decrease. The same applies in magnetics. You can have any electrostatic-charge field strength, separately, you have a separate motomagnetics tensor that want to keep that object as the same strength – whatever that level might be.

Here is a second analogy. If you have a bedsheet laid out on a large flat surface, and you pull one point of it up.

Dynamic #1

    Hand holds up the bedsheet (electrostatic charge)

    Gravity wants to pull the bedsheet down (nucleomagnetics)

##PICTURE

This creates the strength over distance basic formulas – like Coulombs Law for electrostatic charge.

Dynamic #2

    Release the bedsheet and gravity wants to pull the bedsheet down (nucleomagnetics)

    Yet, air pressure wants to keep the bedsheet just where it was (motomagnetics)

##PICTURE

Of course, this is just an analogy, but it gets the idea that we have a fundamental Newtonian balance, and a time-sensitive balance.

> The second part of this dynamic is that changes are limited. This is a distance versus time question. The fundamental was explored in E=mc-squared by Einstein. The field cannot change by more than the speed of light ('c') – distance over time. Therefore, even if forces drive something, there is a limit, and multiple ratio. Therefore, energy from a particle's force cannot express faster than the building or changing of the fields it creates (which is the speed of light).

Fundamental Basics of the 'Electromagnetics-3'

Charge-magnetics is a continuous balance of two very different, yet integrally-linked forces:

- Charge
- Nucleomagnetics

With their distance-and-time-sensitive interaction to the fields of force that they create:

- Motomagnetics (macro-world traditional magnetism)

Charge has spherical orientation for its strength field. That means the field strength is the same in every direction. The Charge force decreases at 1/distance-squared ($\frac{1}{d^2}$). That is covered by Coulomb's Law:

$$F = k\frac{Q_1 Q_2}{d^2}$$

Magnetism is directional with a north-south orientation for its strength field. It decreases at 1/distance-cubed ($\frac{1}{d^3}$) toward poles. The field strength is different at different inclination (longitude) relative to that axis, and the force has complete symmetry at any latitude (azimuth) at the same inclination (longitude).

That field strength shape is stronger at the equatorial direction which is something fundamental to understand.

Those two are different than motomagnetics and its north-south.

However, for our calculations, the orientation will be the minimal force which is the north-south orientation, and so it will follow the 1/distance-cubed ($\frac{1}{d^3}$) formula.

Motomagnetics is the change in the magnetics field. Or better, the tendency for the field to m. As something move, one side is a deficit and the other a surplus; those are north and south, the deficit is attracted to the surplus, yet repulsed by another deficit. Motomagnetics is traditional, observed magnetics.

At the magnetic poles,

$$B = \frac{\mu_0 IA}{2\pi d^3} \text{ where } IA - magnetic\ moment$$

Depending on the angle from that magnetic pole, then

$$B = \frac{\mu_0 IA \sqrt{1 + \cos^2 \theta}}{2\pi d^3}$$

Then the overall Force is a complex calculation in all directions:

$$F = given\ orientation\ complexity, too\ much\ for\ here$$

So, the magnetic force looks like either of the below:

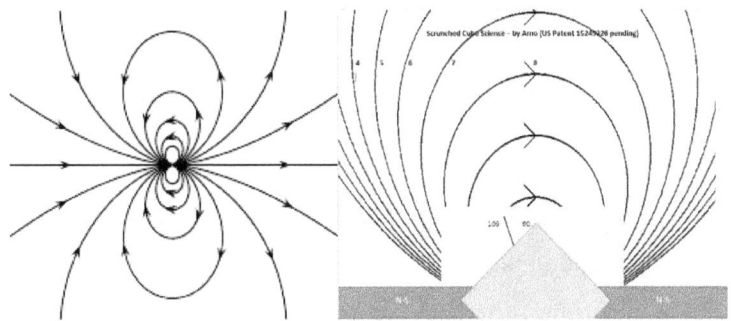

Both operate based upon balancing such that opposites attract, and like types repel. Therefore, for every attraction created, there is a repulsion created depending on what is on the other side of the interaction.

The picture is the same for nucleomagnetics, except there is no direction lines. The proton nucleomagnetics field is attractive to protons, at both poles. The proton nucleomagnetics field is repulsive to electrons, at both poles.

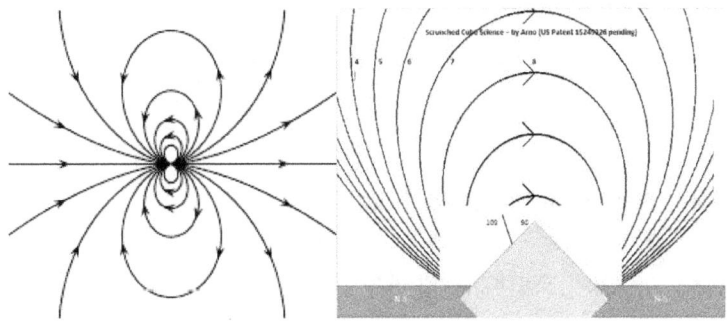

Both Charge Force, Nucleomagnetics, and Motomagnetics Force Balance: Balanced Attraction and Repulsion

Force balances positive and negative charges spherically in the electrostatic charge force:

- o Like charges repel each other:
    - o Positive charges repel positive charges
    - o Negative charges repel negative charges
- o Opposite charges attract each other, truly balancing:
    - o Positive charges attract negative charges
    - o Negative charges repel positive charges

However, those create a Newtonian opposite that is axis-oriented, nucleomagnetics force:

- Proton particles attract protons
- Proton particles repel electrons
- Electron particles repel proton particles
- Electron particles do not attract electron particles (I have not yet determined if they repel or if they are neutral)

As those particles, and thereby or maybe more precisely as <u>their fields</u>, move, or even rotate, the magnetic field changes, which I have called "motomagnetics", for a balanced, directionally north and south for the magnetic force:

- Like magnetic poles repel each other:
    - Positive magnetic poles repel positive charges
    - Negative magnetic poles repel negative charges
- Opposite magnetic poles attract each other, truly balancing:
    - Positive magnetic poles attract negative charges
    - Negative magnetic poles repel positive charges

This second type of magnetics (both with similar based name, '-magnetic' because both express fields with a similar axis-inclination based strength field) is the more commonly understood "magnetics". Hereafter, it is important to discuss them separately to get the attributes applied properly.

This is fantastic. **The fundamental forces, electrostatic charge and nucleomagnetics, and their movement field force of motomagnetics, follows all of Newton 3rd Law (criterion #3).** There is an attraction and a repulsion. Yin and yang.

This give two continuous, stationary functions (criterion #1), if direction-oriented appropriately, of course. Charge (blue) is

1/distance-squared ($\frac{1}{d^2}$), and magnetism (red) is 1/distance cubed ($\frac{1}{d^3}$).

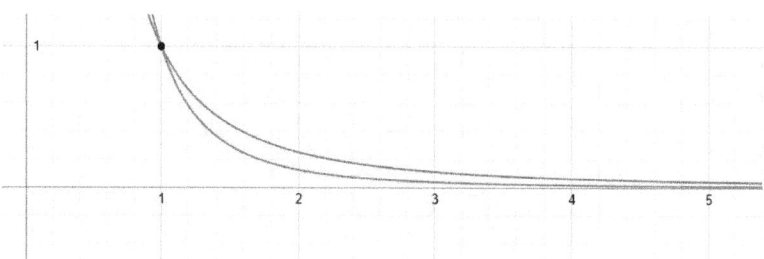

This is more fantastic. **The two charge-nucleomagnetics forces are continuous functions (criterion #1) working over continuous time and continuous space (criterion #4).** Both are stationary fields of permanent particles, continuous and long-standing. The two also meet each of the other criterion. These are the two which are the only ones used (criterion #5). Of course, the basic two describe the first two known forces (criterion #2). Charge is charge, nucleomagnetics is nucleomagnetics, and motomagnetics is derived from the two. Further, charge and magnetism only gets generated from the long-stable particles – protons, neutrons and electrons which also is the criterion #5.

This is the foundation upon which the rest of the postulates shall build.

The rest of the postulates to follow will meet all criteria now if build upon charge and magnetism. Forces and functions that only contain charge and magnetism will meet all five (5) success criteria. I will not need to prove all five at each step. If the source is charge and magnetism, then the solution meets the five criteria.

Settling on a Presentation of Electrostatic Charge, Nucleomagnetics and the Net Charge-Nucleomagnetics Forces

I will present these two forces in a number of ways in both positive and negative. This strength ('y') direction depends on whether the interaction is with a) like charges which is repulsive (positive) above or b), or opposites which are attractive below.

We looked at the two forces to compare the shapes of the resulting strength fields. That graph looked like this:

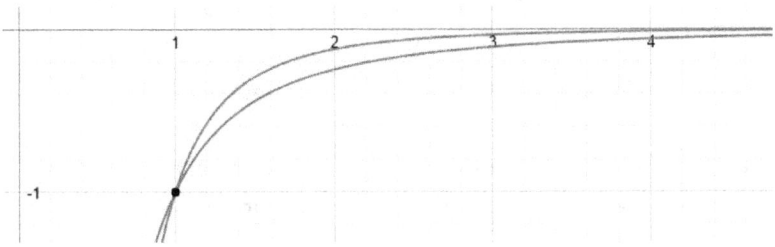

The above shows two forces, but really it missed the point:

Correct graph has those charge and nucleomagnetics functions as Newtonian opposites

45

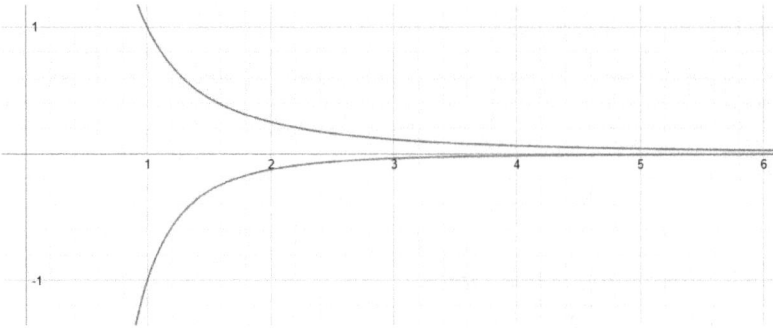

Postulate #1 is that electrostatic charge and magnetism operate in opposite directions in field strength. You will see that graphs hereafter show the fundamental postulate. Both types of force exist and interact in the above, opposite interaction. The charge and magnetism are the same overall strength and opposite field strength direction.

The above display of electrostatic charge opposing nucleomagnetics is the most basic presentation of the fundamental the charge and magnetism are related opposing forces from which all the rest of this postulate flows.

Postulate #2: Electrostatic Charge and Nucleomagnetics are the Newton 3rd Law opposites.

This follows the basic Newton principle, 'for every action there is an equal but opposite reaction.' A charge repulsion has a magnetic attraction of the same field strength. A charge attraction has a magnetic repulsion of the same field strength. A later chapter will actually do that calculation.

Corollary #2A: Charge forces and its activity (movement) exists with corresponding Magnetism effect. One reacts to the other. Always show electrostatic charge and nucleomagnetics with opposite signs in calculations

Remember that the balancing forces can be of somewhat structure. It can be a sphere or a magnet-like north-south – apples and oranges.

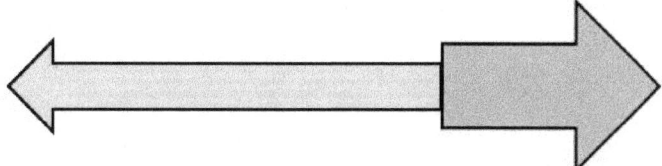

That has one complexity which is that while opposite, the two are not exactly opposite, charge (blue) is 1/distance-squared ($\frac{1}{d^2}$) and spherical, and magnetism (red) is 1/distance cubed ($\frac{1}{d^3}$), relative to the north-south orientation direction, where those two have different shapes.[vii]

In real life, we have these mismatched Newtonian opposites all the time. We push a wall while on roller skates, but we move, but the wall does not seem to move. Yet, the wall and I are apples and oranges. The wall is attached to this huge thing, the Earth. I am not. Therefore, the opposite actions are different types. So, the force of the wall pushing back does equal the force of me pushing. We really push the wall in ratio of our weight versus the weight of the entire world as the wall is solidly connected to the world. The wall moves back a tiny nanometer. Newton works.

It is understanding these apples-and-oranges combinations that leads to every device in physics – from computers/hard drives to rockets which have a force. A rocket is force in every direction, but the shape of the vessel and opening direct that force only away from the direction, the Newtonian reaction, that the engineers want to move to create push to where the engineers want to go, and blocking the other possible directions. We can use these to differentiate actions and reactions to great advantage. The only action from a rocket is out the back, so the rest moves in the Newtonian reaction. Stuff goes out the back, so the structure moves toward the front with the same opposite. Most importantly, the stuff going out the back goes faster, but has less mass. It is a complex calc of not must how much, but also the speed. The balancing is at combined energy level.

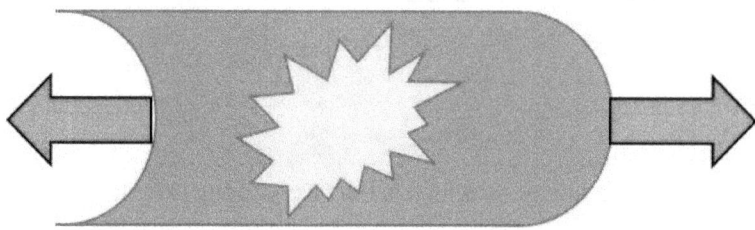

Electrostatic Charge and Nucleomagnetics in Combination – the Basic 'Net' Force

Throughout this adventure, I will focus on simplified visual graphs. That is my style; graphs identify breaks and combinations best for general audience presentations. It does not take a PhD in math to understand when the graph has an understandable shape.

Presentation Guidelines for Force Graphs

1. When shown separately, electrostatic charge will be red, and nucleomagnetics will be blue. When shown together, the various net will be a) orange for strong force, b) green for weak force, and c) purple for both molecular bonding and gravity (since they really are the same thing which I will explain in a later postulate).

2. All force presentations will follow the international standard; that is, positive 'y' values will be repulsive, and negative 'y' values will be attractive.

3. Distance from the measurement particle or nucleus set of particles extends to the right as positive numbers.

4. Similarly, Z = Protons, (+) = Protons, N = Neutrons, e- = Electrons, (-) = Electrons. And, (ZN) = All particles in a Nucleus as the integral sum or combination of Protons and Neutrons (together nucleons).

5. Units for various graphs may described later or separately. For the ease of presentation, often the measurement will be 1, 2, and 3 of an established unit of measure. The 1

might be the Coulomb constant, the Bohr radius, the diameter of a particle, and such base factors. That makes the scale not overwhelm the basic picture.

6. Distance units will be described within the following ranges:

Distance Range Name	Distance from center	Associated Force	Arno Vigen Scrunched Cube (AVSC) Name
Sub-nucleon range	Less than 3 * 10^{-16} m	Sub-atomic particle physics	
Nucleus binding range	Less than 10^{-16} m and 10^{-15} m	Strong interaction force	Binding
Electron shell range	Less than 10^{-13} m and 10^{-12} m	Weak force	Settling
Molecular Bonding Range	Less than 10^{-12} m and 10^{-11} m	Bonding forces	Bonding
Distant Objects	Greater than 10^{-11} m	Gravity	Gravity

Again, a picture probably makes these ranges clearer:

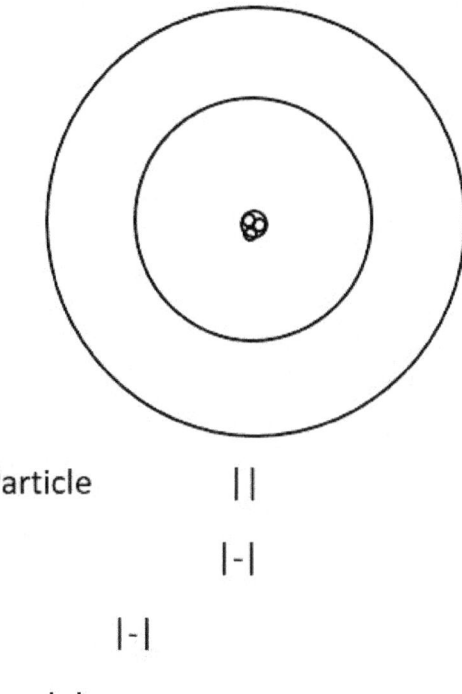

Basic Nucleon Particle ||

Nucleus |-|

Electron Shell |-|

Molecule Bonding |-|

From the particles, you get both charge and magnetics. Charge goes in every direction (X,Y,Z):

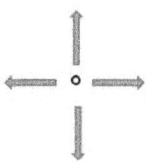

Magnetics goes in a north-south orientation:

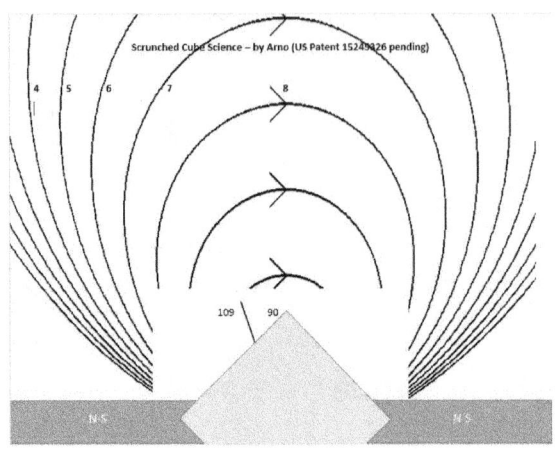

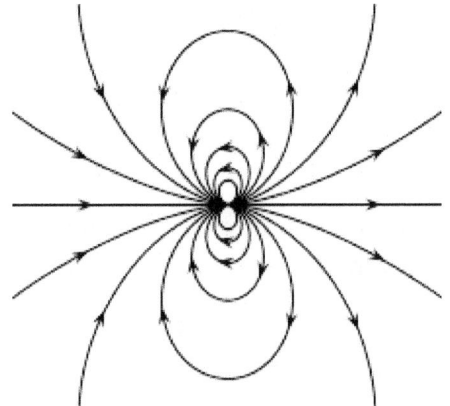

The related 'named' fundamental forces are what happens in the ranges between those natural locations.

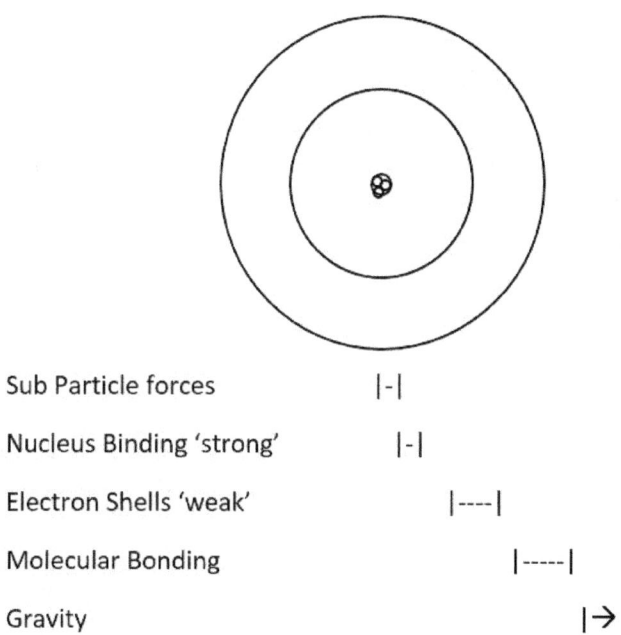

Sub Particle forces	|-|
Nucleus Binding 'strong'	|-|
Electron Shells 'weak'	|----|
Molecular Bonding	|-----|
Gravity	|→

We will look at this in two steps. The fundamental force combination of charge and magnetism for one particle, then the combination of multiple particles and both charge and magnetism within certain distance ranges and geometric configurations that are currently called each of the above-named forces.

So, the other fundamental forces, beyond charge and magnetics, operate in specific distance ranges. Hmm! For centuries, scientists thought the forces as separate, giving them unique names and sometimes their own subatomic particle, but maybe the other fundamental forces are from the same base two, but the distances create the specialty combination of forces and particles that give the attributes we know by each name for that range of distance.

The Fundamental Force: Netting the Force of Charge and the Force of Magnetics

The most basic presentation is the net of the two forces. This is electrostatic charge (red) and nucleomagnetics (blue). Both decrease over distance. However, magnetism decreases faster after the balancing point (> 1), yet that same math makes it increase faster at tiny distances (<1).

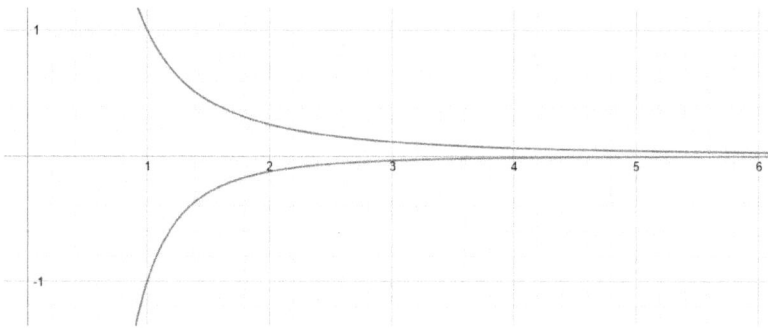

Most importantly, the 1st postulate here is that electrostatic charge and nucleomagnetics are opposite, the 'equal and opposite' but of different types. Whenever charge acts, magnetism has the opposite force.

This graph shows the charge (blue) which is spherical, and the magnetism which for this graph is oriented towards north-south.

All of the secrets to the fundamental forces contain unobserved opposites so that the fundamental forces are stable . . . and follow Newton's Third Law of Motion. The purpose here is to explain each force as the combination given the situation of the range of distance which drives the elementary forces into the know structures.

Where the opposite is not observed, you will find that the view is only charge or only magnetism, and the Newton 3rd expresses as impacts to the field not observed. Look for it.

The Net Charge-Magnetics Force Does Have Distance Difference Profile

While the 1st postulate is that charge and magnetism are opposites, they have the different orientation and profile.

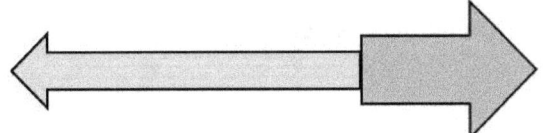

That means that the net force does have a positive and negative range. Both positive and negative so internally offsetting overall.

Net Charge-Magnetics by Distance

For a Single Particle (North-South Direction)

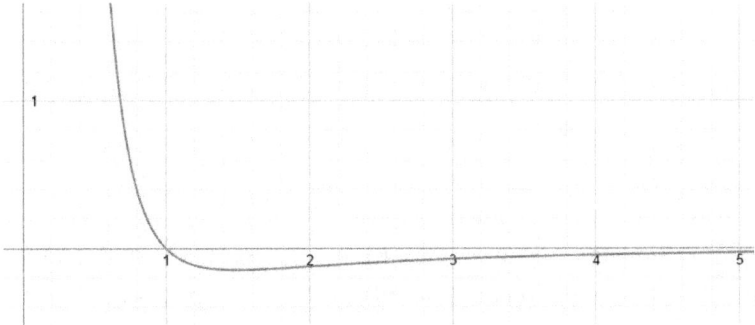

The above is the 1st postulate net force over distance for a single particle (to another single particle). It is the combination as opposites of the charge force and the opposing magnetism force.

At Each End, Charge and Magnetism Mask That the Function is the Combination

The magnetic field effectively decreases quickly in this calculation. Over anything but those smallest atomic distances, scientists only use the charge calculation, and ignore the magnetics because it is decreasing exponentially faster.

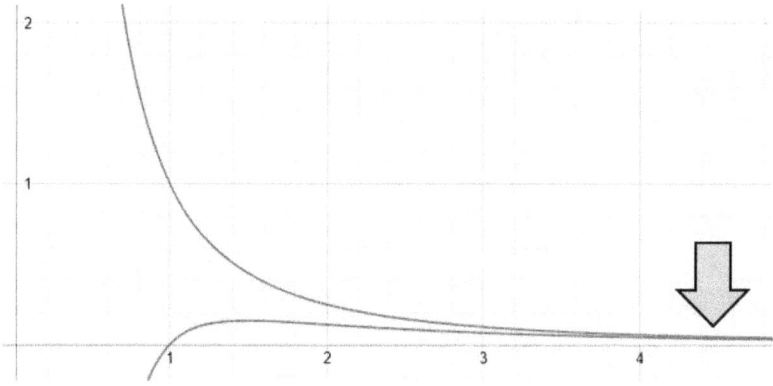

Charge Force (Blue) Versus Net-Charge-Magnetics

You can see that at normal distances ($> 10^{-11}$ m), scientist forget the nucleomagnetics and use the Coulomb's Law of Charge only. The two are effectively identical. We can understand their simplification. However, we must remember that nucleomagnetics was just ignored for insignificance in that particular range.

At the tiny end, near zero, the opposite happens so that the force at close distances resembles more the nucleomagnetics force. Too often, scientist forget magnetics, like the 'strong force' textbook above. In each relevant range, scientists should integrate the magnetism calculation, or combo calcs, for ranges near or inside the dimensions of the shells and bonds.

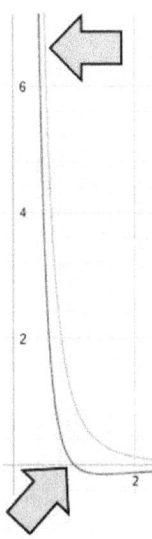

The combo electromagnetic force (red) as small distances is effectively just the magnetic force (orange); magnetism is that strong at the nucleus.

Strangely, the combo force goes negative (attractive) after a point, yet the magnetic force alone is always positive (repulsive).

It is that range in the middle, the electron shell, which needs lots of understanding, and always needs both elements in each calculation. Remember that; it will be critical as we go on.

For some of the amazing formulas, we will focus on two points in the middle in the net charge-magnetics force graph. We will look at the zero and the maximum attraction (or minimum repulsion if you think in reverse). Remember that negatives are attraction!

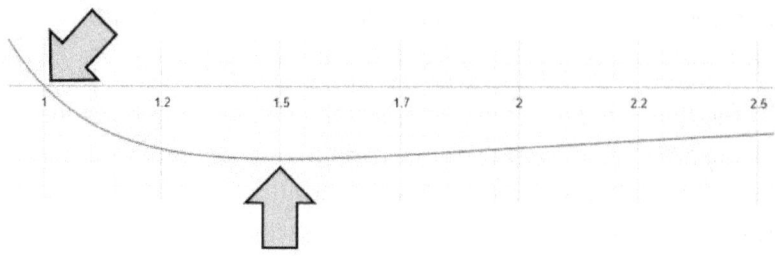

What is that magic point of balance (zero on the graph)? **For our purposes, the distance of 1, is the average radius of the electron shell (R_{ES}).**[viii]

That makes perfect sense. An electron is held in a shell at a natural, consistent point of balance. Everyone keeps searching for a reason why there is the shell distance. However, from the graph above it is obvious. The must be a place of balance between electrostatic charge and nucleomagnetics. And . . . we already know that the nucleus protons and electrons have a balancing structure called the shell.

That distance is about 10^{-12} m. It is the volume enclosed in that shell based upon the force from the integral sum of the constant charge-magnetics field-force of a fundamental particle or grouping of particles (Z=Protons, N=Neutrons) that is a nucleus (ZN)[ix]. For the single proton and single electron 001-H Hydrogen, that radius is the Bohr radius. For one proton, it is easy to calculate. It was named for Niels Bohr, the great scientist that discovered this important point where the "behavior of particles" changes (voila, exactly as the graph shows). Bohr's observation a century ago that a particle (electron) acts differently inside that radius matches the postulate.

Of course, that radius distance for more complex orientation elements gets discussed in detail later. For now, using the Bohr radius as the average will get all the basic calculations close enough. Bohr in his lifetime, and a century of scientist afterwards, have not resolved all of those complexities of larger atoms shell placement. However, using the net charge-magnetics combination, with "scrunched-cube" geometry, does resolve the electron locations and average radius, electromagnetic spectrum, and bond angles for all size atoms, for every element.

In fact, many textbooks show that electrostatic (charge) force as the only known force in those ranges. Note that the 'electrostatic' force shown in the textbook example below[x] is just the charge force shape earlier, not the 'net charge-magnetics' profile in my above chart. Therefore, the below from a textbook has not integrated the known magnetic force component at all – when I know that force exists and that it applies in that distance range.

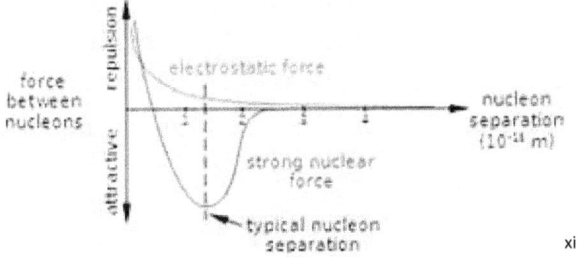

[xi]

Especially fun, it is shown as magically starting and stopping by the non-Newtonian scientists:

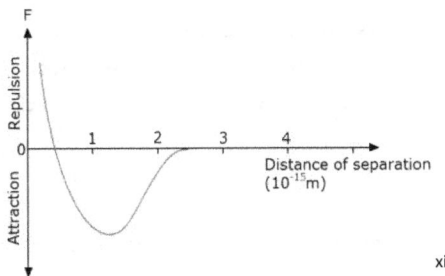

[xii]

[Some advanced explorers probably noticed that the loop goes deeper in the above textbook example than in the basic 'charge-magnetics' graph on the last page. However, we will show why the loop section is bigger for the 'strong nuclear force' in that future chapter. That trick is one of the amazing feats of nature. This teaser gives the adventurous explorer time to start guessing. See if you can noodle the reason for that extra deep dip, the droopy loop, of the force graph before that chapter.]

Beyond that range, at molecular bonding distances, $> 10^{-11}$ m, the magnetic force is $\frac{1}{10^2}$ times (0.01) less than the charge force. From that point on, it is a rounding error, and ignored. You can see this in the basic graph of the charge force (blue) only, and the net of charge minus magnetics (red). The red line is the charge (blue) less the magnetic force which I call net charge-magnetics. At the right, the two lines are indistinguishable.

Postulate #3 – The Balancing of Charge and Magnetism Is Exactly What Creates Electron Shells; Net Charge-Magnetics Is the Weak Force.

The function is just like Charge for distance objects, and just like magnetism as tiny distances. That does not mean that the offsetting elements is not there, but its strength balances focusing at the other end. In most cases, the other expression of charge-magnetics is not observable, and gets dropped as insignificant from calculations.

Note that at the electron shell, both are equally powerful. That will become some magic.

The graph of the weak force is the same graph that we already introduced. It is the most basic combination without any extra particles, special orientation, or integrals.

Weak Force = Net Charge-Magnetics for Opposite-Type Particles

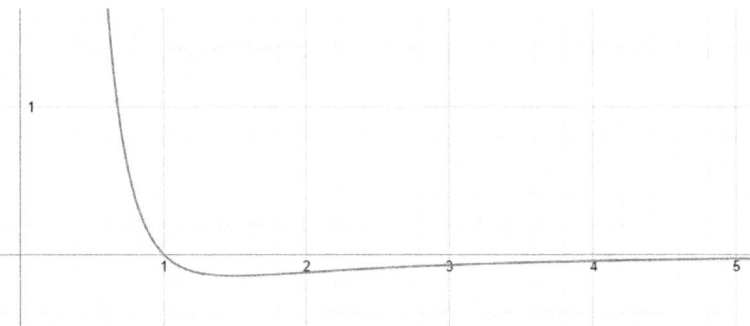

So, for the electron shell field – as an average distance, yes, the three forces add, but two of them exactly offset. The positive charges of the protons in atoms are exactly equal in number, but opposite in Charge than the electrons. Therefore, for the differential geometry integral, those completely offset. That leaves only magnetics in the determination.

Postulate #3-A - That Electron Shell Point of Balance Is the Integral of Magnetic Nucleus Particles (Z,N) Strength Force At the Average Bohr Radius Distance

For an atom,

Protons create attraction (+)	# of Protons (Z)
Electrons create repulsion (-)	# of Electrons = # of Protons (Z)
Magnetic field keeps the electrons apart from the nucleus - repulsion	Based on protons plus neutrons (Z+N)

So, the calculation of the net field strength is simple:

+Charge(Z)

-Charge(Z)   which equals the charge of its electrons

+M(Z+N)

Total   +M(Z+N) as the Charge of Z cancels out

Leaving only the magnetics of the nucleus particles (Z= Protons + N=Neutrons).

The first two offset for the differential geometry vector analysis. This is a huge math trick so you don't have to calculate the location of every electron in electron shells. Worse, you don't have to calculate the electron-to-electron repulsions for every permutation of electron combinations. [For a fuller presentation on actual electron shell placements see Electron Shell Chemistry is Just . . . Scrunched Cube Geometry, Book #2 of that Simple Words to Understand . . . Chemistry series on the shape and orientation of the electron-shells.] For electron shells, you actually do calculate the location of each electron in 3D within that directional magnetic field. But, for the average needed for just fundamental forces like gravity, you can skip those complicated steps entirely.

Yes, there is minor adjustments:

- The distance changes by the electron location at angles to north-south because the magnetic field is north-south gets stronger at $\sqrt{1+f(\theta)}$, which my preliminary calculations limit to up to $\sqrt{2}$ at 90 degrees, towards the equator.

- The distance changes by speed (velocity) of the electrons in this type of orbit which then might have a relationship to temperature (heat energy). Refer to angular momentum for the adjustment calculation.

- The force calculation has an extra difference because $\frac{1}{d^2}$ calculates at an extra measure of force when the d varies versus the average. The integral of random (stochastic) $\frac{1}{d^2}$ is not $\frac{1}{d^2}$ of the average d of that random set. The integral is greater than the average calculation. These adjustments are the complex math of quantum mechanics.

These adjustments are real, but not critical to these postulates. We will need those adjustments eventually for exact solutions, but the adjustments do not change the fundamental force from charge-magnetics which are the 'big picture'. So, back to 'big picture'.

Fundamental Charge-Magnetics Force: The Net

Repulsive at Close Distances and Attractive at Large Distances

In most cases, this set of forces, charge netting magnetism would be attractive. Sometimes the other particle is the same (repulsive above), and sometimes the other particle is the opposite, so attractive below.

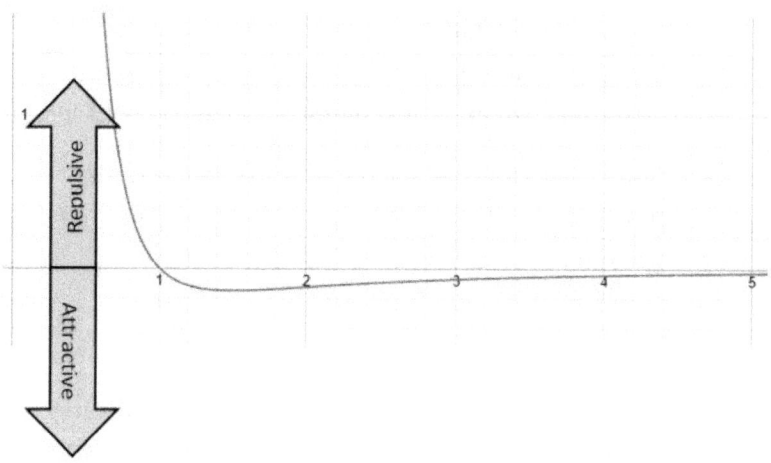

For example, nearer to zero you have a proton near a proton, which is repulsive (+ on graphs). We know that fact from nuclear explosions that when protons get near protons, the proton react and repel with fierce power. This graph reflects this standard feature. That force graph reflects common sense reality.

Further, at long distances, we have a net force that is a) attractive (>3) decreasing at 1/distance-squared, just like gravity. Gravity is the name of this force for distance objects (with a tweak we will discuss in that chapter). Again, the same 'net charge-magnetics force' matches common sense observations.

NOTE: The gravity calculation includes the net of more than two particles, the positive (+) nucleus protons offset by the negative (-) electron shell distance, to get the scale correct, which will get introduced in that chapter. For now, the direction and shape for the function are perfect which makes the final step easy.

Therefore, the direction and scaling of this fundamental force graph also connects with reality at both ends of the spectrum.

In the middle, there is a balance, and common sense tells us that we need balance to create electron shells.

At all distances, the postulate charge-magnetics formula generates results matching the common-sense observations.

Fundamental Charge-Magnetic Force: The Net Viewed for Opposite Force – Charge versus Magnetics

Yet, remember that any one of these force graphs can get flipped depending if you are looking at the charge expression of the combined forces or the magnetic expression of the combination.

- This force when looking at an opposite type looks like:

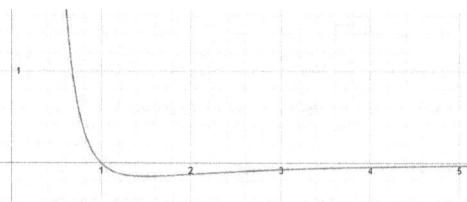

- This same force when looking at a like-charge flips:

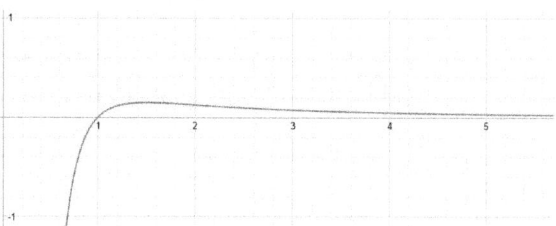

Nuclear decay/explosions and gravity are charge based interactions. The observed effects are what happens to the same-charge particles at close range. If you look what is happening to the magnetic field – the reaction – you see the reaction in magnetics. When a nuclear explosion occurs, you have an EMP pulse. Guess what, that EMP pulse works on the reaction function curve – the magnetics. That is why it is called EMP – electro-<u>magnetic</u> pulse.

Therefore, if looking at magnetic forces only, the observation might be:

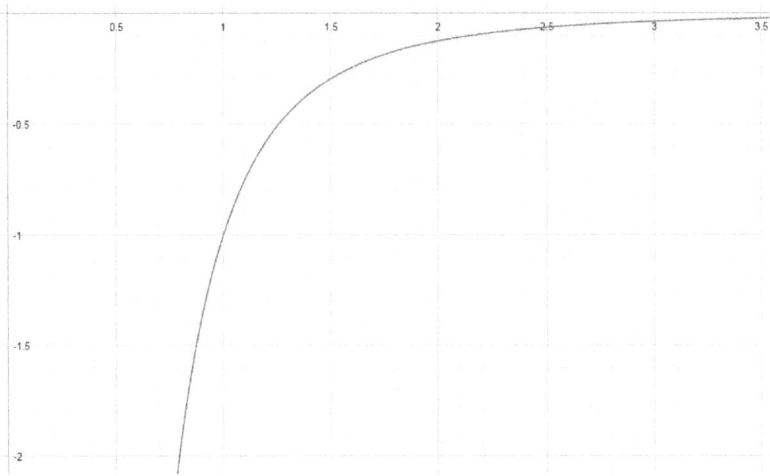

There is a huge 'pulse' as the magnetics alone, or charge alone, it exponentially more powerful.

This type of fundamental occurs when you have the same charge, positive-to-positive, or negative-to-negative, or north-to-north or south-to-south. So, don't forget these net charge-magnetics express in both forms, and as charge or magnetics alone where appropriate.

The above graph displaced is the basic function of charge-magnetism. It will be the basis of every force derived hereafter. Charge and magnetism are fully linked, but express as Newtonian opposites.

Charge and magnetism are balanced overall, but that balance has an important differential, the fundamental graph above, where the balance is repulsive close offsets attraction at distance, or vice-a-versa depending on the particles in the interaction.

Graph of the Charge-Magnetics of a Proton

The force for a proton when looking at an opposite type looks like:

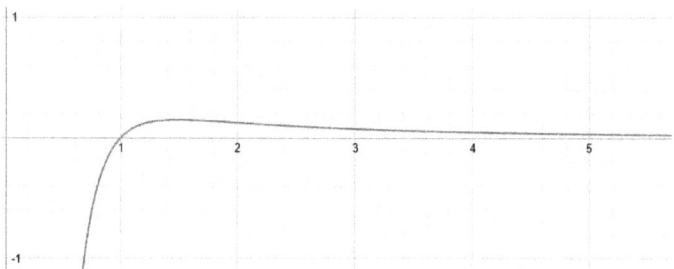

Graph of the Charge-Magnetics of an Electron

The force of an electron when looking at an opposite-charge flips:

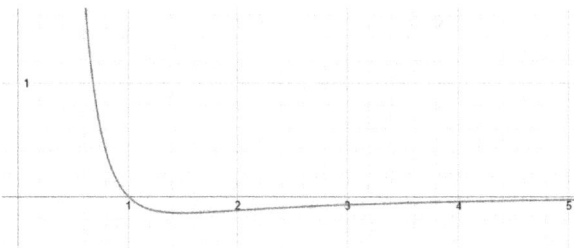

Graph of the Charge-Nucleomagnetics of a Neutron Leads to Strong Force that Holds Every Nucleus Together

However, the force for a neutron has the charge elements cancel. That is why it is called a neutron. A neutron:

- Has no net electrostatic charge
- Has nucleomagnetics

Therefore, we only have a force function that looks like the magnetic force curve (the red) only.[xiii]

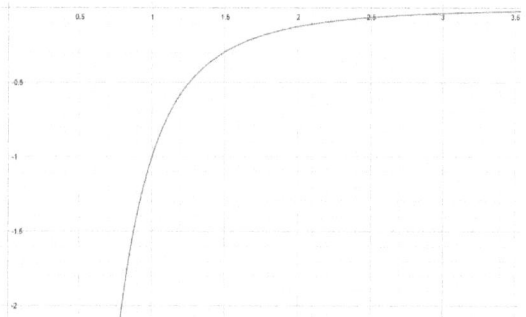

Note that the neutron graph 'y' is always attractive (negative on force graphs). This makes it very different than protons and electrons. And, guess what neutrons do; they only attract protons into nucleus structures.

Further, notice that the strong attraction occurs closer to particle. Again, that matches our common sense. We know neutron because they 'stick' (are attractive) in a nucleus. The neutron's work occurs at the left, very powerful part of the force function.

Yet, there is a trick about magnets. They remain strong when connected in a chain. That is, the magnetic strength is a flat line until the physical connect ends. Therefore, the force comparison graph goes from:

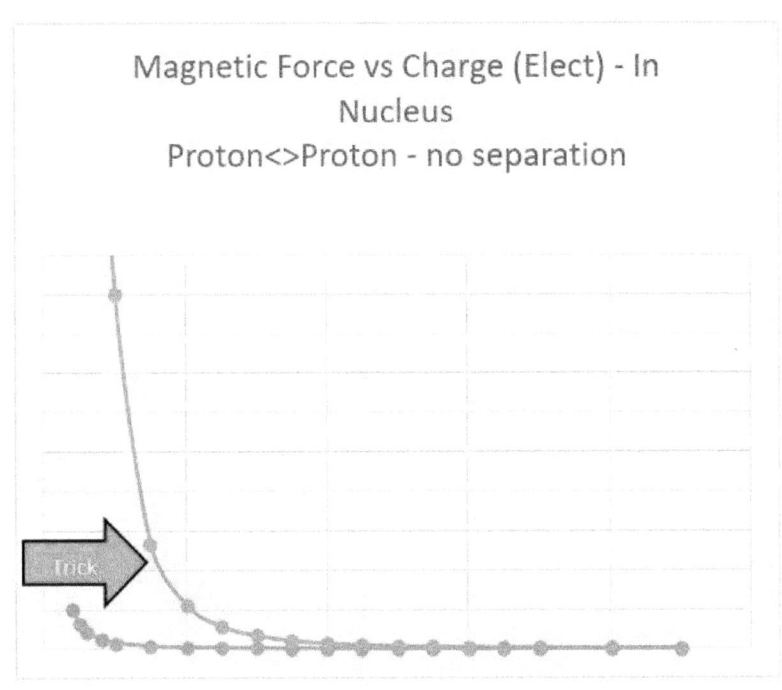

To

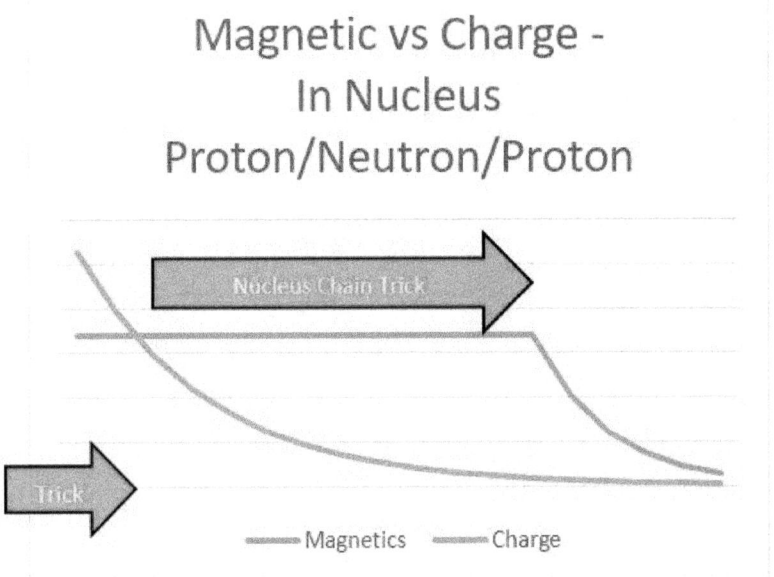

Magnetic Chains Proton-Neutron-Proton Strong Force

I already said, "Magnetics is more powerful inside the electron shells." Ah! There is the secret. In this range, particle binding follows magnetics more than charge. The charge repulsion of proton-to-proton can get overwhelmed by magnetics.

In this structure, proton-neutron-proton, the magnetic force remains strong because it is a chain. The charge repulsion (proton to proton) drops because of the stable separation of the neutron.

Normally, the magnetic force drops faster ($\frac{1}{d^3}$ versus $\frac{1}{d^2}$), **but** magnetics forces stay strong in structures (chains and rings), so the magnetics, in a chain, will keep a nucleus together, even with the proton-proton repulsion force.

Magnetics Stay Strong in Chains

Postulate #4 – A Nucleus Physical Structure, Proton-Neutron-Proton, in a Magnetic Chain or Chain-Ring or Combination Has Magnetics That Stays Strong, and Enough Distance Between Protons So That Proton-Proton Charge Repulsion Has Decreased. To Remain Stable, a Nucleus Structure Must Have At Least One Neutron As Physical Separation Between All Protons. The Entire Magnetic Structure Is Stable As Magnetics Is Stronger Than Proton Charge Repulsion At Those Distances.

One of the main qualities of magnetics different than charge is north-south orientation. Further, that orientation has another attribute. The strength of the magnetics fields continues at full strength until the source, the structure, stops.

In a bar magnet, you can see that strength all the way to end of the bar. Further, you can see that the field operates as if just one magnet. It skips the connection point (after a tiny distance within the tolerances that we need). The entire structure looks like one large magnetic at distances greater than the magnetic length.

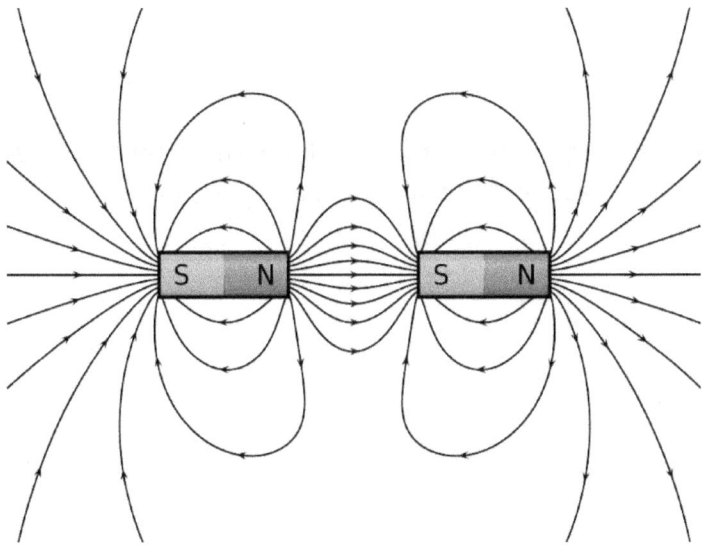

For the postulate to work, at the levels of the nucleus particles, the magnetic-force must exceed the charge-force.

Note that I am taking the force curve as continuous. The same charge and magnetics formulas that apply at the Bohr radius apply at the nucleus particle distance.

Both parts of this 'strong force' formula have only charge and/or magnetics.

The underlying function uses both charge and magnetics for the protons. It is based upon the net of charge and its Newtonian reaction magnetics in the opposite direction (so all 5 success criteria apply) with the net force already discussed:

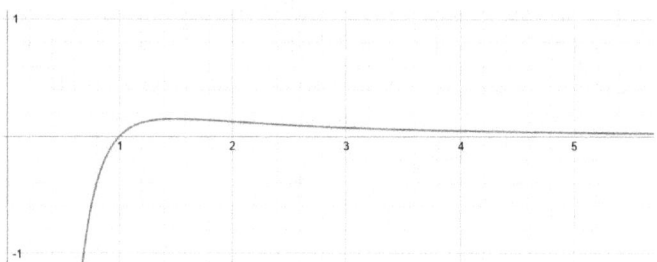

The underlying function is the same for the neutrons only uses magnetic components (so all 5 success criteria apply):

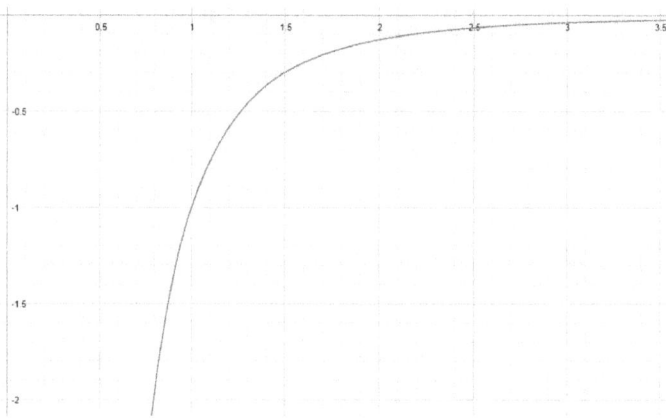

The strong force is just those two charge-magnetics for the special situation where for a particular proton, a neutron separates it from the next proton. We can graph these with the two forces, but applied at the different distances. That means we have two particles in the graph of 'strong interaction', not just one (like 'weak' force).

Of course, given the two are equal at the Bohr radius, and the magnetism will inverse at cube versus the square of charge-force, one would expect that magnetism is must stronger at the distance that of nucleus particles.

That distance would be 2 x radius since the magnetics must transfers via the one neutron particle in order to attach to the 2nd (a proton).

Understanding the Magic Balancing Point

At first, I used a place marked as Point 1, where charge and magnetics balance. There must be such a point of balance if their strength follows different structures.

Using the balancing point where forces balance simplifies lots of work. At that balancing point, the forces are the same. Instead of 1, we can use the charge energy from Coulomb, kQ_1Q_2, or a magnetic energy $M(Z,N)_1 \int \theta_1 \ M(Z,N)_2 \int \theta_2$.

Either of those are the same at the magic distance (R_{ES}) and its volume of magnetic energy:

$$Volume = \frac{4}{3}\pi R_{ES}^3$$

Since the forces balance, we have this amazing equivalency. We can substitute either depending on the challenge to get the easiest solution.[xiv] Of course, that is limited to complete systems. Complete is stable, balanced atoms. And that means atoms with the same number of protons as electrons.

1) That balancing distance should get used to show the field that drives the particular specific force for that situation.

2) At that point, we may also find something fundamental about the relationship of charge and magnetism. It might open understanding of what makes them interact.

How to Make Radius of Electron Shell into a Constant

Most people would consider the radius of an electron a variable distance (R_{ES}) because:

1) In larger elements, there are multiple shells which means that even within the same element, the distance varies.

2) In different elements, the distance of the electron placement is different. The evidence from electromagnetic spectrum analysis is overwhelming. Different real colors are the distance between various shells.

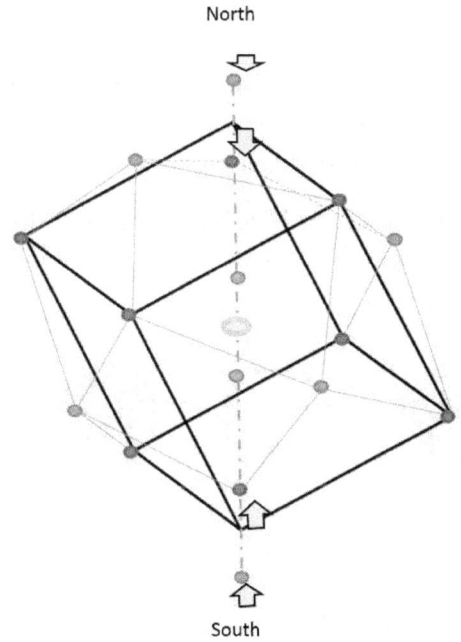

Yet, this is a balance point, and it is balancing magnetics. The magnetics are a fixed quantity from each nucleus particle, a proton or a neutron.

If there is the charge energy of XX protons pulling the electrons in, there must be XX in magnetic energy pushing them out.[xv]

Postulate #5: The Volume of the Electron Shell has a Direct Correlation to the Number of Magnetic Nucleus Particles. (With very minor adjustments for a) internal cancellations for the nucleus chain-ring shape, structure called 'mass deficit' or b) for any the magnetics of electrons not contributing because they sit between two nuclei in bonding 'bonding mass reduction')

As each nucleus particle adds to the nucleus, it comes with a block or blob of magnetic energy. These basically add (except some loss is the structure wrap around):

Generally, the magnetics add like adding blobs of the same energy. That creates a bigger blob having the same total strength (volume):

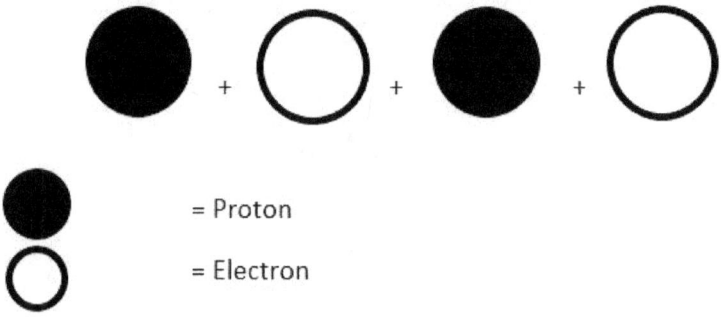

= Proton

= Electron

Although for large nucleus, the magnetics might build to overlap or interfere:

Or, in a ring, actually a portion would have a small portion of the field in one of 3D (X, but not Y, or Z) that cancel each other out (creating what is called 'mass deficit'):

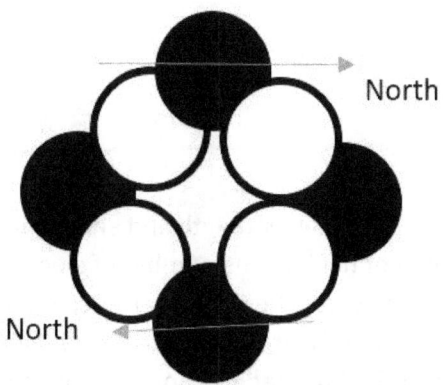

* In nucleomagnetics, I it really not 'north-south', until motion introduced, but the concept that the field flows around is valid.

Remember that expresses in the other dimensions, the 'z' axis perpendicular to the plane of the nucleus ring, the magnetic fields for those same particles would <u>not</u> cancel. In the other directions, the magnetic fields act in the same direction – so no cancellation from that portion. In that sense, the cancellation is only partial.

So, for some of the magnetics calcs, elements with more particles, the total magnetic force is slightly less than the particles alone, or if the structure was just in magnetics-in-a-line.

Comparison to Bohr

Magic Point 1 contains field strength and energy, in Gauss fashion, the basic strength of field for the number of electrons (Z,N) * volume of a base magnetic field energy from a fundamental particle. For the 001-H Hydrogen electron, that volume of the sphere created at the Bohr radius. Bohr was on to something fundamental. At this distance, the two forces, charge and magnetics are the same, but in opposition, and therefore balancing.

Niels Bohr determined the point of balance for a simple 001-H Hydrogen electron. His experiment determined the point at which external charges had an impact on 001-H Hydrogen electrons. Inside, charge stopped its primacy. Inside that distance, the atom focused on the internal, not external. The Charge force stopped being the primary force.

Bohr did not connect that Newtonian opposite magnetics gain priority, but he knew that point was important, and related to fundamentals like the Planck constant and the speed of light.

Now, Niels Bohr and others failed to determine this point of balance for more complex atoms. For them, because they didn't check magnetics, the location where charge loses its primacy. Their attempts failed because **the Bohr calculations did not take into account the 3D geometry (the magnetic axis symmetry) of the magnetic components of the fundamental charge-magnetics force combination**. So, nothing worked for Bohr and many others after him beyond the basic calculation for 001-H Hydrogen atoms. Bohr did not have the fundamental 2-factor double equation. Bohr did not consider magnetics and 3D.

That point, 1 on my scale, is where the charge and magnetics balance on a straight line towards an electron from nucleus. On the outside, (>1) yet, charge is more important. Inside that radius, something was different. Bohr and others recognized that the environment of the atom did not act the same at that border location.

So, given that we are dealing with two interconnected forces as the fundamental charge-magnetics force, we find that those two have different strengths at difference distances from a single particle.

Those distances, and the fundamental charge-magnetic force preference is as follows:

Distance Range Name	Distance from center	Primary Force	Limitation
Sub nucleon	<3.0E-15m	Magnetics only	Physical basic particle size
Nucleus binding range	3.0E-15m to 1.5E-14m	Magnetics, then Charge	Still room for charge repulsion if protons-to-proton distance so charge still important, stable when created as the magnetic chain (Proton-Neutron-Proton-Neutron) or as ring structure for higher elements
Electron shell range	1.5E-14m to 1.0E-12m	Balance of charge and magnetics	Individual Particle Repulsions creates shell structure 2,8,8,18, and such based upon north-south filling in 3D structures.
Molecular Bonding range	1.0E-12m to 6.5E-12m	Charge as primary, yet some magnetic structure	Directions allow some attraction and some repulsion
Distance Objects	> 6.5E-12m	Charge as primary	Integral (averaging) gives tiny net attraction

To complete this postulate, an explorer must address each challenge with a separate priority based upon calculating the distance applicable to that situation.

By the way, (ZN) is the integral sum of number of nucleus particles, or more specifically, the integral of their magnetic force in the nucleus as configured. That is something concrete. We can find and count the number of nucleons; we don't need 'spin' or 'color'. We do not need another quality called 'mass'. The answer is already in the Period Chart for every element.

Magnetic Field Is Stronger Inside the Electron Shell, Charge Stronger Outside the Electron Shell

To finish this postulate, an explorer must release the common understanding that 'Charge is the strongest force'. That is true in our world, the visible, but we don't view the world inside the electron shell of an atom.

However, a better understanding of chemistry and molecular physics refines that understanding into:

- Magnetic Fields are Stronger Inside the Electron Shell

- Charge is Stronger Outside the Electron Shell

Fundamental Question: Why Doesn't Proton Charge Repulsion Make Nucleus Fall Apart?

In a range of 0.5 to 5 times the nucleus particle size, the first interesting combination occurs. At this level, a proton will repel a proton. Yet, we observe a stable nucleus for millions on years. Only over such long times, a portion of something like C-14 degrade into C-12, do decay; some particles (neutrons) deteriorate. Similarly, Uranium and Plutonium have observable breakdown of nucleus, **but for all those breakdown, there is the vast majority of nucleus which stay in some very large complex structures for long period – thousands of years.**

In this range, the nucleus range, the magnetic force is generally greater than the charge force.

The priority of charge and magnetics apply:

- At and beyond the shell range, charge is more important because 1/distance-cube ($\frac{1}{d^3}$) magnetic decrease to insignificance.

- Inside that boundary, magnetics is more important, as the strength of that field is larger than the charge force – *at the same distance*

In addition to a flip from attractive to repulsive or repulsive to attractive, a full understanding includes that the importance of magnetic force changes to priority within the $(ZN)* V_{R_{ES}}$.

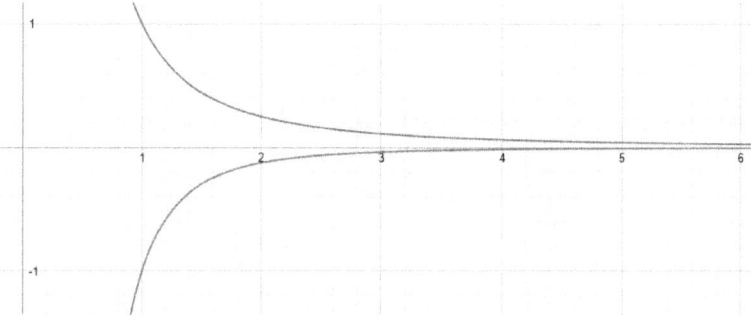

To the right of 1 (>1), the magnetics (red) line is closer (smaller) than the charge (blue). To the left of 1 (<1), the magnetics (red) line is bigger (smaller) than the charge (blue). The red is sloped more. So, the net line flips.

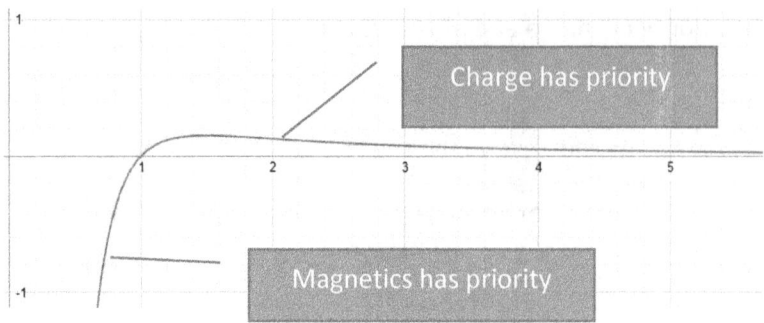

Yes, charge and magnetics are opposite. That is based upon the basic Newton Law "equal and opposite". Everything must come in some sort of balance. For everyplace that charge has priority, there is another place where magnetics has a balancing priority. Part is positive; part is negative. Yin and Yang. Newton would be happy.

If you look at the force separately, like Bohr and others, you would say the function is discontinuous. I review of charge would start at 1 and be nothing closer. Magnetics would disappear at 1.

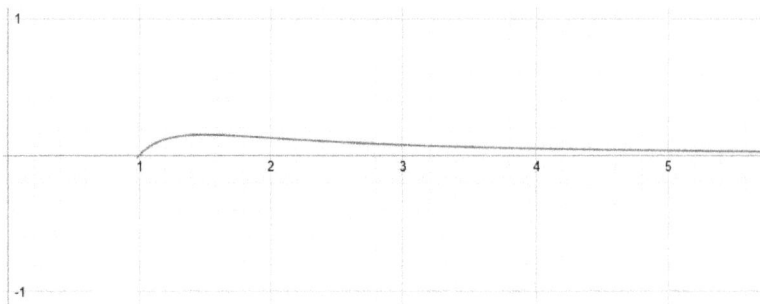

The above is what Bohr observed. Charge force started at 1, it did not act like Charge (Coulomb) inside that radius. This acceptance is one of the ways that scientists accepted discontinuities. It is OK to have something start magically at a distance where there is no concrete.

Remember that the above graph is correct. It works for ranges starting a little outside the electron shell. I cannot refute it – in that range. Bohr was amazing and brilliant in his discovery.

The challenge often is that researchers observe charge, but rarely observe magnetics. Magnetics is very difficult because:

a) It is oriented, so if you are at a different angle, the measurement is different. Therefore, no one gets consistency in observed results. Hence, the tendency towards statistical solutions versus physical explanations.

b) This level of magnetics is magnetics of protons and neutrons and such. We don't have indicator that can measure magnetics in such small ranges. *We don't have a compass only nanometers in size.*

c) Magnetics of whole atoms is done in stable configuration which are a combination of particles which creates more complexities into the measurement. One cannot observe magnetics of a single proton 001-H Hydrogen, the simplest atom. Just think what would happen with more complete Carbon C-12 or C-14 or the overloaded Uranium which probably has protons and neutrons within the nucleus generating fields in every magnetic orientation. As a result, measuring magnetics of the core nucleus particles is nearly impossible.

It is given that magnetics as tiny distances is critically important, yet impossible to measure for the last century. However, we have a new base postulate of the strength curve of magnetics and the combined charge-magnetics. We can use that to test the #5 postulate.

Measuring Charge-Force versus Magnetic-Force at Distance in Nucleus

Let's test this #5 postulate (nucleus proton-neutron-proton physical structure) for proton-proton charge-force repulsion at a tiny separation and then for proton-neutron magnetic-force attraction at that separation.

It will compare three particles, forces and distances:

1) The Charge Force at a Distance of Two Particles (with separation)

2) The Magnetic Force at that same Distance

3) The Charge Force at a distance less than one Particle (without separation)

The hypothesis is that #2 will be larger than #1, but not larger than #3. That is a nucleus can build proton-neutron-proton, but it cannot build proton-proton. The magnetic (green) force should exceed the repulsion (yellow) force.

OK

Repulses as charge-force gets much stronger without the neutron separation. If 1/2 the distance, the strength is 1/4 more strong.

Binding Fails

Nuclear Decay

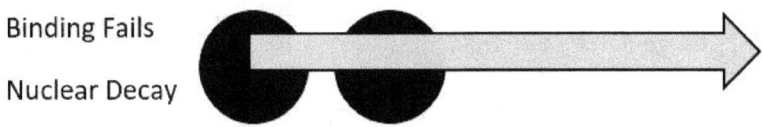

The charge force at a nucleus distance would be:

$$F = k \frac{Q_1 Q_2}{d^2}$$

Factor	Charge-Force Estimation
Charge-Force Constant (Coulomb)	$k_e = 10^{10}$ [8.99×10^9 m³ kg²/(s²)]
Charge 001-H Hydrogen atom = 1 proton	$q_1 = 10^{-19}$ [1.67×10^{-19}]
Charge 001-H Hydrogen atom = 1 proton	$q_2 = 10^{-19}$ [1.67×10^{-19}]
Distance	$d = 0.86 \times 10^{-15}$ m or 10^{-15} m
Exponent shortcut	+k+Q+Q-d-d
Short-cut calculation	10-19-19-(-15)-(-15)= 10-19-19+15+15= 10-38+30=+2 N (Newtons)
Charge-Force Repulsion	10^{+2} m¹/(s²) (Newtons)

Yet that must be less than the magnetic repulsion at that same distance – which it is.

$$F = M_A \frac{n_1 n_2}{d^3}$$

Factors	Magnetic-Force Estimation
Charge-Force Constant (Coulomb)	$M_A = 10^{-38}$ m³ kg²/(s²)]
Charge 001-H Hydrogen atom	n=1 or 10^0
Charge 001-H Hydrogen atom	n=1 or 10^0
Distance	$d = 0.9 \times 10^{-15}$ m or 10^{-15} m
Exponent shortcut	+k+Q+Q-d-d-d
Short-cut calculation	-38+0+0-(-15)-(-15)-(-15)= -38+0+0+15+15+15= -38+45=+7 N (Newtons)
Magnetic-Force Attraction	10^{+7} m¹/(s²) (Newtons)

So, magnetic-force is larger compared to charge-force at the same nucleus particle. That is, if the particles are separated by the distance of a particle.

However, the real comparison is that charge-force, if proton-near-proton must also create the repulsion seen in nuclear decay and nuclear reactions. If no separating particle, the protons get much closer as shown below. The proton-proton repulsion is huge, even versus the magnetic attraction.

$$F = k\frac{Q_1 Q_2}{d^2}$$

Factor	Charge-Force Estimation
Charge-Force Constant (Coulomb)	$k_e = 10^{10}$ [8.99×10^9 m³ kg²/ (s²)]
Charge Proton	$q_1 = 10^{-15}$ [0.9×10^{-15}]
Charge Proton	$q_2 = 10^{-15}$ [0.9×10^{-15}]
Distance	$d = 3.0 \times 10^{-16}$ m or 10^{-16} m
Exponent shortcut	+k+Q+Q-d-d
Short-cut calculation	10-16-16-(-10)-(-10)= 10-16-16+15+15= 10-32+30=8 N (Newtons)
Charge-Force Repulsion	10^{+8} m¹ / (s²) (Newtons)

If there is not the neutron separation, then the charge force can get too close, and become 10^{+8} (~10,000,000) m¹ / (s²) vs the magnetic attraction of 10^{+7} (~1,000,000) m¹ / (s²) of proton-proton.

If there is the neutron separation, then the charge force is overcome because the 10^{+2} (~100) m¹ / (s²) charge-force when

91

separated by a neutron is then less than the magnetic attraction of 10^{+7} m^1 / (s^2) of proton-proton.

As such, the forces at play do not bond proton-proton, but do bond proton-neutron-proton.

Defining Particle Forces Versus Combo-Forces

We have the 'net-charge-magnetics' already that looks like:

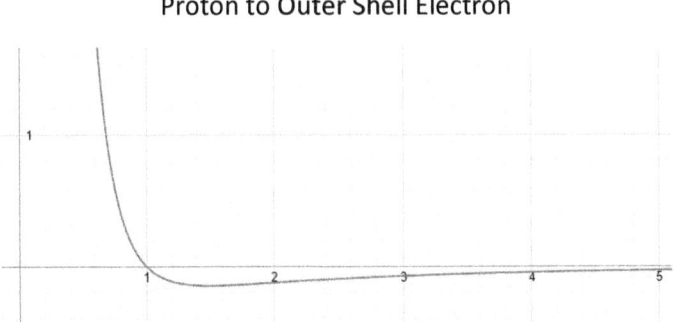

Proton to Outer Shell Electron

This is the force on a location based upon the distance from one particle. However, this is a combination of charge force and magnetic force. As such, it has a change of direction (at 1.5 above). A single force cannot change strengths.

In the above, the line is the net force of a proton in the direction of that proton on an electron. It is attractive at distances outside the electron shell which matches common sense.

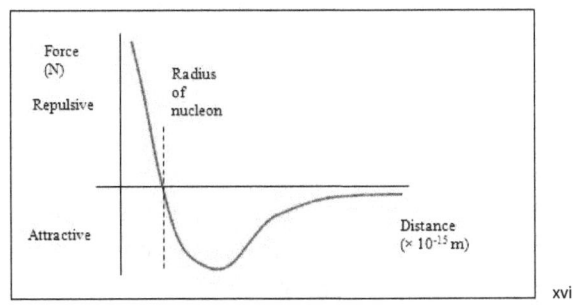

xvi

So, the loop is a little deeper. What would make that happen?

Yet, it you look at the below, you see the yellow graph with something like the shape of the strong force.

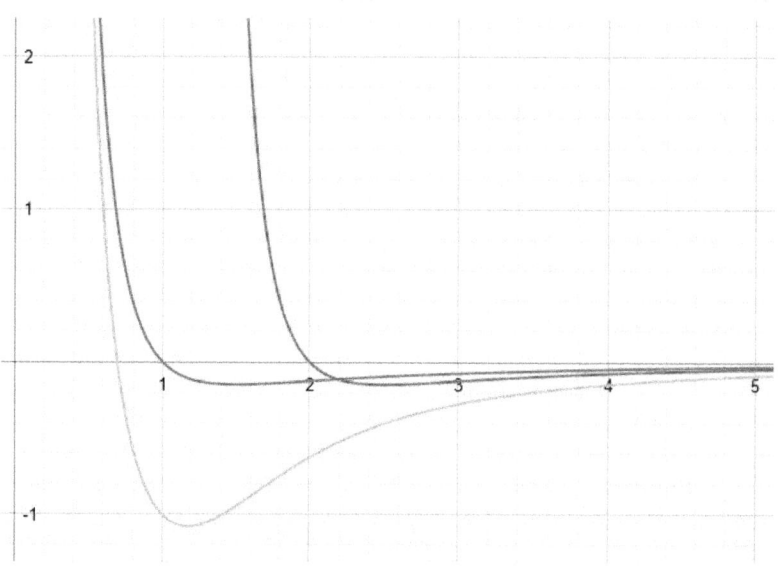

■ $x^{-3} - x^{-2}$

■ $(x-1)^{-3}$
$\quad - (x-1)^{-2}$

■ $x^{-3} - x^{-2}$
$\quad - \dfrac{1}{(x-1)^2 + 01}$

There it is, that extra deeper loop (orange) versus the 1-particle charge-magnetic force. It looks just like current textbooks description of 'strong interaction'.

So, what is the graph? It matches the shape of the strong interaction/force. The graph consists of a standard Proton magnetics (10^{-3}) less the reaction/Newton opposite proton charge (10^{-2}) with a neutron (no charge, but only magnetics) between. However, that in-between is expanded because the chain proton-neutron-proton so the decrease does not start until the 3rd particle. That is seen in that decrease does not start until the distance of the 3rd particle (2nd proton).

Remember that the picture is about the energy to separate the 3rd particle.

Again, a picture will help:

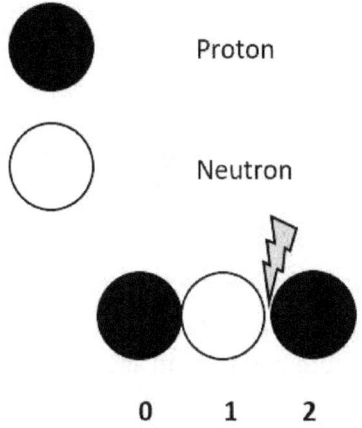

0 1 2

But really the distances are from the 'surface' for math purposes

0   1

The calculation is the force required to remove the Proton #2 (a minus sign in the graph, negative are attraction), but from two different particles – a) Proton #1 has charge-magnetics, and b) Neutron #1 which is just magnetics.

$\dfrac{1}{x^2}$	Proton0 Charge	Proton Charge starts decreasing from the neutron at 1
$\dfrac{1}{x^3}$	Proton0 Magnetics	Proton Charge stays stable from the Proton at 0 to the Neutron at 1, then decreases
n/a	Neutron1 Charge	N/A – A neutron carriers a net zero charge.
$\dfrac{1}{(x-1)^3 + 1}$ *	Neutron1 Magnetics	Proton Charge stays stable from the Proton at 0 to the Neutron at 1, then decreases

* that +1 takes into account that this extends the chain so it is averaged

So, the combination of a proton net charge-magnetics with a neutron, a) which is magnetics only, b) as the separation, provides a force that match the observed 'strong force'.

Now actual results have various angles how the P-N-P-N-P might bend to form a ring. In turn, that angle changes the size of the budge. So, an explorer could delve more deeply than allowed at this level.

1) This requires the forces from two particles. This is obvious because it has a slope of zero.

2) Where is the 2nd particle? It is sitting at 1 from the 1st particle in the direction of the force (N-S) and 1 to the side.

3) Why does the slope change at the location of the particle plus a little 1.2 vs 1.0?

This is the graph of a central proton to a 'shell orbit' electron. It has the formula which adds the charge forces and the magnetics forces.

$$F_n = \sum_{k=0}^{n-1} (F_{Q_n} + F_{M_n})$$

In words, the force on particle n, the last particle in a system, is the sum of the individual forces, both Charge (Q) and Magnetism (M) from each of the other particles.

It is easy to identify a single force; it has no place that has slope = 0. A place that move from negative to positive or visa-versa.

The below Charge and Magnetics do not change direction in strength.

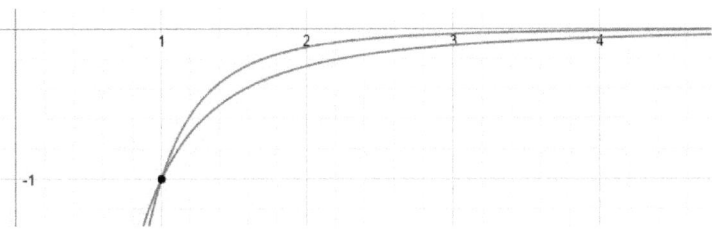

Yet, as we have seen, the combination does – but only once. So, a point of slope = 0 does not necessarily mean there is a new particle at that location. It often means that this is a reflection of two apples-oranges forces from the one particle.

Corollary #5A: Strong Interaction Force and a Gluon is Really a Combo Force of the Combination Closer Magnetics-Only of a Neutron and the Net Charge-Magnetics of a Proton Separated by the Neutron, but Connected by Magnetics, so a Stable Structure

Since the 'strong force' is the interaction of two particles, the gluon is not a particle itself.

The reason scientists thought about a gluon, as a particle:

- The force starts at its own distance making people thinks an additional particle exists at that distance

- Scientists did not integrate the magnetics force into the calculation, especially the net-charge-magnetics, that causes (start) of force at the distance that magnetics overwhelms even charge at the distance that particles touch.

Gravity As the Charge-Magnetics Attraction Force of Electrons Being a Little Closer than That Repulsion from Protons.

Charge attraction is governed by Coulomb's Law.

$$F = k \frac{Q_1 Q_2}{d^2}$$

q = charge of each molecule, and

k = the charge-strength constant.

Please focus that Charge attraction decreases at 1/distance-squared. This is similar to the current definition of the Gravity force.

$$F = G \frac{m_1 m_2}{d^2}, \quad G = 6.67 \times 10^{-11}$$

Remember that part earlier where I said that for distance objects, the magnetics gets ignored, and it truly insignificant. Gravity definitely follows that. Both force graphs have that shape – for distance objects.

For distance objects, magnetics is not material to the calc.

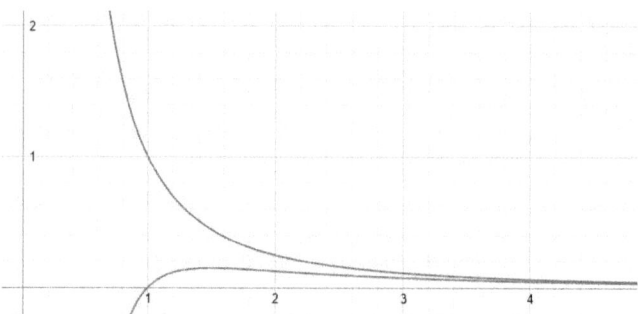

That common $\frac{1}{d^2}$ at the right of the force graph is the key, most important place to start. It makes sense that Charge-Force and Gravity-Force are directly related. Both decrease at the same rate. The evidence is hugely documented and proven that both charge and gravity act in 360 degrees, even spherically, in exactly the same ways.

1/Distance-Squared ($\mathbf{1/d^2}$) goes forever and decreases exponentially

1/distance-squared in a formula means **a) that forces goes on forever** (and that the distant object looks at that force from one of those directions). The force gets smaller, but never disappears. Ah, like gravity! Charge (and gravity) never disappear. They keep working even a galaxy away. The square also means that **b) the change in distance reduces the force exponentially.** A 1/distance ($\frac{1}{d}$) calculation for a distance of 3 is three times (3x) as far, but 1/distance-squared ($\frac{1}{d^2}$) is not three times (3x) smaller, but 9x smaller. **Squared is exponentially more powerful.** Remember that a) and b) *double fact* for later.[xvii]

The challenge* is connecting the other parts Gm_1m_2 and kQ_1Q_2 to each other. m_1 is mass, G is Newton's gravity constant, and kQ_1Q_2 where k is the charge-force constant and q charge particles.

* That challenge is significant because current attributes of charge (q_1) do not match well with mass (m_1). Charge is the # protons, and mass is <u>the sum</u> of the # of protons and the # of neutrons. Except for Hydrogen, those are very different base numbers.

Electron-shells is Another Part of the Equation

The placement of electrons in shells creates that difference in Charge force that drives the net-force, 2nd part, of the equation.

But, on the surface, this part of the theorem, that electrons are closer, seems challenging. An 'orbit' is both closer and further away. So, doesn't that cancel also? The back ½ cancels the front?

No, the close and far don't exactly cancel. A 1/distance-squared (*remember squared is exponentially more powerful*) force creates more than enough extra pull on the close side. The attraction at the close side of the orbit locations exceeds a) the repulsion from the protons in the nucleus; even with b) low attraction when the electrons are on the other side. That is the nature of 1/distance-squared.

A simplified example will explain why the nucleus-shell structure with electrons in 'orbit' generates a 'tiny' net force at 1/-distance squared:

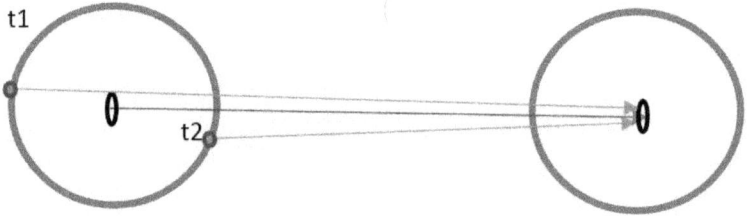

For simplicity[xviii]: compare the force which acts using the distance-squared (as is the case for electromagnetic Charge and gravity) when the nucleus sits at a simple 10, say meters, (dropping all zero exponents/E-XX) and the electron moves in an orbit of 1.

That makes the nucleus each sit at 10 apart, and the electrons orbits at a distance from 11 at time1 and 9 at time2. The system moves back and forth 11 then 9, then 11, then 9.

Interaction / Time	Force / Distance-Squared	Calculation	Subtotal (in Force units)	%
Proton<>Proton				
Time 1	$-1.0 / 10^2$	0.01000		
Time 2	$-1.0 / 10^2$	0.01000		
- Total			-0.02000	100%
Electron<>Proton				
Time 1	$1.0 / 9^2$	0.01234		
Time 2	$1.0 / 11^2$	0.00826		
- Total			+0.02060	103%
Grand Total			+0.00060	3%

So, whenever there is a nucleus-shell configuration, charge **must** create some net-over-time (as they orbit) force towards distant objects. It is not 3% because my example numbers do not reflect the actual shell-distance difference ratio, but there is some 'tiny' extra-net charge force that has not been included in previously calculations trying to determine gravity.

That net-charge goes on forever. That is the nature of 1/distance-squared. It has all the common properties of charge **AND** gravity.

1/distance-squared makes closer items more powerful

Obviously, we need to add a few zero's to get the correct calculation. However, any person can understand the basic concept of the above closer-electrons average-over-time 1/distance-squared calculation. There must be a 'tiny' extra force when two opposite Charges exist in a structure of nucleus and orbits. With the electrons pushed out, the outer force is net a little closer for any $1/d^2$ calculation which makes them slightly stronger, and that 'tiny' net-force is gravity.

'Big' Charge versus Magnetics versus 'Tiny' Gravity

Without going into details, you can see that atom-gravity is tiny compared to the Charge Force. For two 001-H Hydrogen atoms at 1-meter apart Charge of 10^{-28} is ~1,000,000,000,000,000,000,000,000,000,000x more powerful than the 10^{-62} force for gravity. Charge is nearer to 100x (10^{-2}) bigger than Magnetism at a meter apart. Gravity is a 'tiny' force when compared to a) the electrical charge of an atom or b) the magnetic field of the nucleus.

Factor for 001-H <> 001-H atoms@1m	Force Calculation	
Electrical Charge Force factor	10^{-28} m^1 / (s^2)	*
Magnetic Force factor	10^{-30} m^1 / (s^2)	*
Established Newton Gravity Force-Earth	10^{-62} m^1 / (s^2)	*

See chapter 9 for specific calculation.

Common sense tells you this order, even if the units are different and beyond human comprehension.

In the smallest spaces of our world, if you have ever gotten a shock from an electrical cord or static electricity, then you know a very small, almost-invisible electrical charge is powerful. A tiny electric cord short-circuit can create visible reactions. Gravity has never given you that sort of shock from something that small. If you have every tried to pry apart two magnets connected north to south, you know that magnetic force is strong, especially those tiny Bucky balls. Yet, again, no shocks from magnetic balls. Compare this to the atom-gravity force holding atoms of water into a bubble on a flat surface; one touch and the 'tiny' atom-gravity force of the bubble gets burst. Atom-gravity falls apart easily. Atom-gravity is tiny.

For gravity to really impact your life, it must be a huge boulder accelerating over a long distance. Of course, even then, a lightning bolt is more explosive. In the big world, charge is still much bigger than gravity.

At this point, what my fellow explorer should understand is that for a nucleus-shell structure, there must be a tiny net-force of the electron attraction. And by tiny, I am thinking of how small 10^{-62} is versus 10^{-28}, so we are probably on the interesting path.

Next, we will move to figure out a) what causes the reflection, then b) how to calculate it, then c) check that the new method of calculating matches Newton.

Postulate #7: Net of Charge of Electrons in the Electron Shell as Integral Over Time Versus the Charge of Protons in Nucleus is the Gravity Fundamental Force

Atom-Gravity

Factor	Base	Field Strength decreases by	Radius
Nucleus creates magnetic field force	Protons + Neutrons = 'Nucleus Mass'	While complex to calculate specific electron balance points within a magnetic field[xix], the total field balance is simple: **(Magnetic Repulsion = Charge Attraction) over / Volume of (Shell-Radius)³**	Nucleus
Electrons has closer Charge force	Net Charge	$$\int \left(\frac{kQ_E}{(d \mp \cos(\theta))^3} - \frac{kQ_P}{d^3}\right)$$ Field Strength Excess Charge of Electrons less the Charge of Protons averaged over time over the electron shell	Electron Shell**
Distant body recognizes only the 2nd larger element $\frac{1}{2\pi} = \frac{1}{\sqrt{a^2-b^2}}$	Product with the Electron Shell's Charge as the 'math eliminated' connection integral in one direction	The Charge Attraction cancels to ***: $$\frac{\int (G_A N_1) \int (G_A N_1)}{d^2 (\frac{8}{3}\pi R_{ES}{}^3)}$$ Which is the same as Newton $$\frac{Gm_1 m_2}{d^2}$$	Galaxies away

* I use Q versus $q_1 q_2$ because this is the fundamental of one Charge attraction or repulsion to one other Charge. It is not a variable, but 1:1.

** The electron shell distance is the combination of the electron-electron repulsion (spherical) plus the nucleus magnetic field (north-south oriented). However, for electron repulsion, there is the exactly balancing number of proton attraction, so in the end for the average the driving element is just the nucleus magnetic field.[xx]

*** The integral is because when a nucleus goes into larger elements, the structure can be a chain-ring where a portion of the ring actual cancels the force observed ('mass deficit')

Gravity by Arno

Nucleus particles magnetism cause electron-shell-distance

Nucleus Particle Magnetism Caused

Gravity is a relayed force.

Relayed

Gravity is a net force. The charge-at-a-distance force of electrons is greater than a) Charge-at-a-distance force of the nucleus protons, even with b) lower charge of electrons on the far side of any electron-orbits -- both because of 1/distance-squared.

Net-Charge- $\frac{1}{d^2}$ Force

Gravity is an integral over time. It is not a separate particle.

Derived *integral over time*

Gravity is a nucleus-particle-magnetism-caused, electron-shell-relayed, net-charge-$\frac{1}{d^2}$-force calculated by an integral over time.

Calculation of Gravity from Charge versus Newton $6.674 \times 10^{-11} \frac{m^3}{kg(s^2)}$

These Electron-Shell Separation has many factors which will get handled extensively in the footnotes. There is a lot of moving parts, many of which need further validation for a 'published paper' level of acceptance. However, in general, you can see these implications of these factors in the below example of 001-H Hydrogen to Hydrogen at 1-meter apart:

- The Charge Force (if not offset by proton) is significant - 10^{-28}
- The Gravity Force (using Newton) is 'tiny' - 10^{-62}
- The Net-Charge Force (Arno method) is a similar 'tiny' - 10^{-62}

The Gross-Charge Calculation for 001-H Hydrogen electron (-) to 001-H Hydrogen (+) atom at a distance of 1-meter is as follows:

Factor	Net-Charge Calculation
Charge Force factor	$k = 10^{10}$ $m^2 / (s^2)$ [9.03]$\times 10^{+9}$]
Charge of orbiting Electron	$Q=10^{-19}$ [1.602]$\times 10^{-19}$]
Charge of distance Proton	$Q=10^{-19}$ [1.602]$\times 10^{-19}$]
Distance	d=1 m or 10^0
Exponent shortcut	+k+Q+Q-d-d
Gross Charge Short-cut calculation	10-19-19-0-0 = -28
Gross Charge Force	10^{-28} $m^1 / (s^2)$

The Net-Charge Calculation for 001-H Hydrogen atom to 001-H Hydrogen atom at a distance of 1-meter is as follows:

Factor	Net-Charge Calculation
Charge Force factor	$k = 10^{10}$ m² / (s²) [9.03)×10⁻⁹]
Charge of orbiting Electron	$Q=10^{-19}$ [1.602)×10⁻¹⁹]
Charge of distance Proton	$Q=10^{-19}$ [1.602)×10⁻¹⁹]
Distance	d=1 m
Radius of the Nucleus (protons and neutrons)	r=10⁻¹¹ m (Bohr radius which is 5.27 x 10⁻¹¹)
Exponent shortcut	+k+Q+Q-d-d
Gross Charge Short-cut calculation	10-19-19-0-0 = -28
Net Electron-Shell Charge Short-cut calculation	-28-11-11-11-1 = -61
Net-Charge Force of the Electron (protons and neutrons) New Formula	10^{-61} m¹ / (s²) /8/3π = 10^{-62} m¹ / (s²)

This is just the basic force of charge force as pushed out by the electron-shell. The gross charge of the electrons 10^{-28} gets reduced by the $\frac{4}{3}\pi R_{es}^{3}$, which is another 10^{-34} creating a net-charge for distant objects of 10^{-62}.

How does that compare with the standard calculation of gravity using Newton's formula?

Gravity from 001-H Hydrogen atom to 001-H Hydrogen atom at a distance of 1-meter using the Newton method is:

Factor	Gravity Estimation
Gravity Constant (Newton)	$G = 10^{-10}$ m^3 kg^2/ (s^2) [6.67)×10^{-11}]
Mass of close 001-H Hydrogen	m=10^{-26} kg [1.602)×10^{-24}]
Mass distant 001-H Hydrogen	m=10^{-26} kg [1.602)×10^{-24}]
Distance	d=1 m or 10^0
Exponent shortcut	+G+m+m-d-d
Short-cut calculation	-10-26-26+0+0=-62+0=-62
Newton Gravity Force	10^{-62} m^1 / (s^2)

The new calculation gets to the same range of strength; net-charge at electrons shells matches Newton's gravity force for a simple 001-H at 10^{-62} m^1 / (s^2).

Net-Charge Force	10^{-62} m^1 / (s^2)

Gravity Using the Graphics of Charge-Magnetics

Once again, let us return to the basic Charge-Magnetics Graph. It comes back to the net Charge-Magnetic Force, yet the gravity force is reduced to that Different of the two types of particles. And those two, protons-in-nucleus versus electrons-in-shells have opposite impact of the distant particles, and thereby atoms.

The two particle charges of protons (red) and electron shell (blue) are very similar. The only difference is that one, the electrons', starts slightly further out. That is, closer to the distant object. However, that create a net (green) which has the same function shape ($\frac{1}{d^2}$), but the scale is reduced by that volume of the separation.

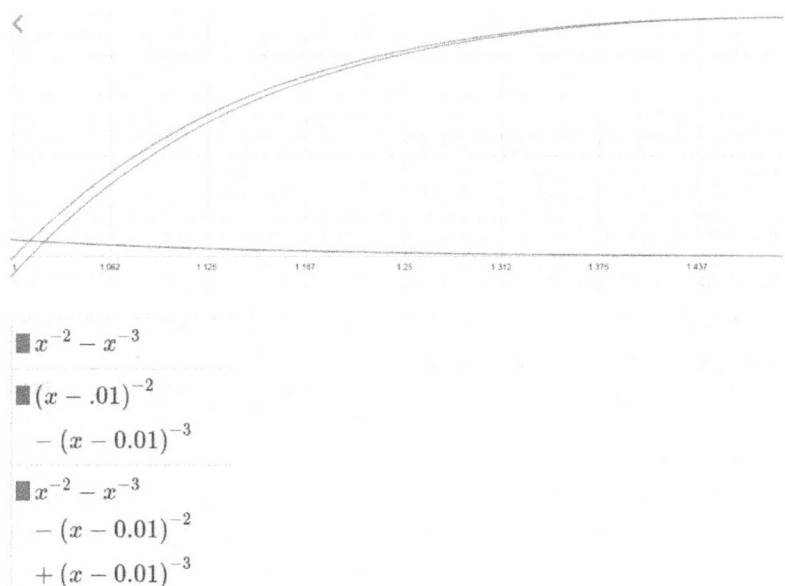

■ $x^{-2} - x^{-3}$

■ $(x - .01)^{-2}$
 $- (x - 0.01)^{-3}$

■ $x^{-2} - x^{-3}$
 $- (x - 0.01)^{-2}$
 $+ (x - 0.01)^{-3}$

Because magnetics decreases more quickly, the gravity net looks just like a charge line – only at a scale much smaller than the full charge force.

At the far right, the shape of all the curves look the same $(\frac{1}{d^2})$. Because it is at the right, only charge drives the shape of the curve. However, since there is a start location different, there will always be a 'net-force' of gravity.

At distance, charge, charge-magnetics and gravity all decrease by $(\frac{1}{d^2})$, so the difference decrease by $(\frac{1}{d^2})$. That is exactly as Newton determined, and experiments reveled accurately.

Electron Shell Balancing Weak Force Important to Gravity

Now that we introduced the extra factor (R_{ES}), how can we make that a fundamental. For what factor is R_{ES} consistent in ratio. This is another amazing magic.

According to differential geometry mathematics, for example Gaussian random field calculations[xxi], the strength of a field is the sum of the field of force elements enclosed. Once you get outside the space (sphere) where all forces are included, then that space is minimized based upon a simple sum of all those force vectors.

Here is the magic. The charge forces that are enclosed at the edge of the electron shell offset. You have X number of protons, and X number of electrons, in a stable atom, they equal, but their charge runs opposite. Thereby, you have the following:

Particle	Contains	Charge Strength	Charge-magnetic Strength
Proton	(+) Charge + Magnetics	+#P	+#P
Neutrons	Magnetics only	None	+#N
Electrons	(-) Charge + Magnetics	-#P (because P = N)	???
Total		Zero	Hmmm! (N+P)

Within that field, the charge balances, so there is only magnetics. Further, magnetics field is a push to the electrons (inside the electron shell). Electrons get pushed away from the nucleus by magnetics – both north and south.

Postulate #7: Electrons, while attracted to charge, are repelled by Magnetics, Both North and South.

The basic understanding of the particles is incomplete. Each particle has both a charge as well as a magnetic property. We teach only the charge properties.

Particle	Charge Property	Magnetic Property
Proton	(+) Negative Charge	North-South
Neutrons	Zero	North-South
Electrons	(-) Negative Charge	***Repulsed by both North and by South***

Science had forgotten Newton. They thought that enough of a system that had the opposites with North-South. However, that would lead to the world locking up over time into a huge chain of particles, north-to-south, with nothing to balance. Nothing would stop everything from building into one atom. However, we don't see that in the sun or stars. The evidence is that something helps separate the magnetics. That is why it is called a shell. It is a layer of particles that protect. The magnetic field fearing electrons are pushed away from the magnetic nucleus. In turn, those same electrons push away approaching other potential magnetic particles. This is why atoms are so stable.

Common sense supports the electrons flee the nucleus magnetic fields. Every electric motor is the rotation of a magnet which sends the electrons away along a wire. That electron fleeing a very powerful.

The priority of charge and magnetics apply:

- o Outside that boundary, charge is more important because 1/distance-cube magnetic decrease to insignificance.

- Inside that boundary, magnetics is more important

In addition to a flip from attractive to repulsive or repulsive to attractive, a full understanding includes that the importance of magnetic force changes to priority within the $(ZN)* V_{R_{ES}}$.

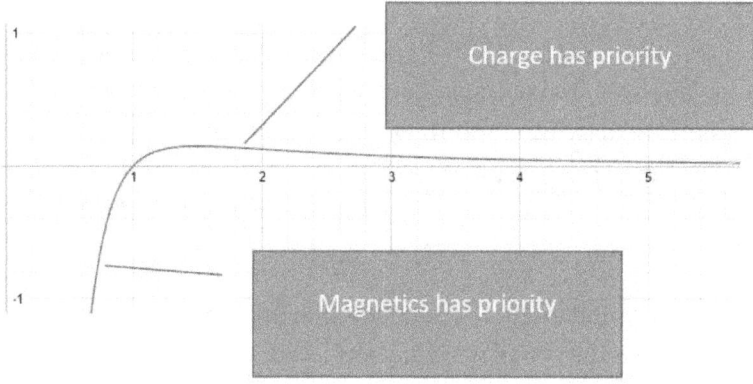

Yes, charge and magnetics are opposite. That is based upon the basic Newton Law "equal and opposite". Everything must come in some sort of balance. For everyplace that charge has priority, there is another place where magnetics has a balancing priority.

Corollary #7A: Electron Repulsion to the Magnetic Field of Nucleus Particles Is the Weak Force

Aha! We have solved one of the fundamental forces using only magnetism. However, magnetism needed a better definition.

Question. What is the strength of that Repulsion?

That was solved years ago. We know that for 001-H Hydrogen, Neils Bohr determined the average electron distance. We know the push is the Charge force, and the electron stays generally in one place because of consistent electromagnetism spectrum analysis including in depth analysis such as the Lamb shift.

That means as Distance = 1, the magnetic force equals the electron-magnetic-repulsion force.

In a volume of space of $(NP)_1$, number of nucleus particles in an atom labeled 1, times volume of balanced charge field and magnetic field (which for 001-Hydrogen is basically the Bohr Radius) ($M(NP)_1 / \left(\frac{4}{3}\right) \pi \, (R_{ES})^3$), electrons will sit in a shell. (R_{ES}) being the Radius of electron shell

Since this is fundamental, a fundamental unit of force from a particle, now the problems with calculating the larger atoms. You can even see hints of this in Niels Bohr notes. He used both mass and the speed of light (c) and 4π.

Postulate #2 Revisited: The charge force and magnetic force balance each other as Newtonian opposites at the average volume or distance of the Electron Shell (R_{ES}).

This balancing will become interesting. Bohr already observed certain effects. If we keep digging at that point, we will find that all the established constants, are one for that distance adjusted for the situation. We can find powerful evidence of the definitive relationship of charge and magnetism.

Most Fundamental Constants Will Consolidate with a Factor Directly from the Average Volume or Distance of the Electron Shell (R_{ES}).

Oh! And as Bohr already proved, that distance is directly related to the speed of light and fundamental quantum of energy, Planck's constant or reduced Planck ($\hbar$).

$$a_0 = \frac{4\pi \varepsilon_0 \hbar^2}{m_e e^2} = \frac{\hbar}{m_e c \alpha} = R_{ES}$$

Yet, that is related to the charge-magnetics force at that distance. Remember that both the magnetics and the charge strength is the same at that distance.

Plus, we know that mass comes from the related at the shell distance:

$$\frac{M(NP)_1 M(NP)_2}{(d^2)} = \frac{kQ_1 Q_2}{d^2} \text{ at } d = R_{ES}$$

When combined with Newton's gravity:

$$F = \frac{Gm_1 m_2}{d^2} = \frac{\int (M(ZN)_1) \int (M(ZN)_2)}{d^2 (\frac{8}{3}\pi R_{ES}^3)}$$

We can solve for:

$$m_1 \text{ instead of } m_e$$

Or as a ratio of the two. We solve and substitute for the radius of electron constant.

$$a_0 = \frac{4\pi \varepsilon_0 \hbar^2}{m_e e^2} = \frac{\hbar}{m_e c \alpha} = R_{ES}$$

But that Bohr standard was created with the elements of the Planck Constant and the speed of light. Yet, to work it needed this 'magic' permeability of space (ε_0) factor. So, the permeability of space (ε_0) is related to the fundamental relationship of charge-magnetics. The permeability of space (ε_0) by the method gets related to the fundamental magnetic force of nucleus particles.

That gives a glimpse into how magnetism is the charge separation and/or rotation and/or field differential at the edges. That full calculation will take another volume, and is beyond the scope of the fundamental forces. That will be Book #3 in this series.

Fundamental Question: This Nucleus Distance Is a Significant Link Between Charge and Magnetics Direct Interaction and Creation

Here is the part that I believe exposes something more fundamental about Charge and Magnetics.

Challenge: The Balance in Strength is NOT at the Edge of the Particle

The easiest path would be a simple application of Newton. You have a particle, and the charge and magnetics field are exactly the same at edge of the particle. The only difference would be that charge goes spherical, and magnetics has the north-south orientation and thereby 'bagel' field strength.

However, that is not the case, the strength balance the charge versus the magnetics-minimum at the Bohr radius.

Fact: At the Particle Edge, Magnetics are Stronger

The fact that magnetics are stronger than charge at the particle edge gives pause.

Fact: At the Particle Edge, Magnetics are Stronger By the Exact Factor of the Charge/Magnetics

The amazing revelation comes from the nucleus binding calculation. The distance that particle separate is exactly the distance where magnetics become primary. That must be something fundamental.

$$\iiint_{x=d_{ZN},y,z}^{x=\infty,y,z} k_Q \frac{Q(ZN_1)}{x^3} = \iiint_{x=d_{ZN},y,z}^{x=\infty,y,z} \frac{M(ZN_1)(1+\cos(\theta))}{x^3}$$

The question is the start of the integral. What is the physical limit which is the start of the integral. The solution to this should make the forces equal, and Newtonian opposite, and establish a true dimensions of the charge-magnetic particle size.

This distance is amazing.

However, if you take one more calculation, you find that the magnetic field strength has a relationship to charge separated by this same distance.

I believe this is one of the fundamental relationships.

"

The charge cannot exist without creating a magnetism that overpowers it as this special distance. The balancing point where magnetism at that separation become larger than charge at any distance is the fundamental dimension of charge-magnetics. Likely that is the dimension of the underlying magnet which the charge inherits. Strangely enough, that distance happens to be the same size as particles (protons and neutrons).

"

Either:

1) A charge has a physical dimension, and as it spins it is constantly creating magnetic fields based upon this radius dimension.

Or

2) A charge is always moving, and the distance is a fundamental of the speed-of-light that magnetism is create in that charges are always moving relative to something, so magnetic always exists as the counter-force (reaction) to charge.

I don't have the answer for this yet. Start your next thesis exploring it.

Molecular Bonding Force

In a range of 2.5 to 6.5 time the Electron Shell, there various directions.

1) One direction has one or more electrons, and that creates a shell of repulsion for other atoms that come closer than 2.5.

2) Another direction has a direct path to the nucleus, and the electrons are at an angle. Because of the angle and that half of electrons are on the other side, there is a net attraction for

That is the where bonding occurs. It occurs in the direction where that differential is greatest. It settles between the electrons, or at location where the outermost shell has room to shift.

1) It creates stable angle based upon the number, and thereby structure of each element. That angle is based upon the building up of electron placements.

2) It creates a stable distance where the other atom likes to sit. Remember that the other atom has its own proton nucleus, so the total structure cannot fit all the way to the distance of the outer electron shell.

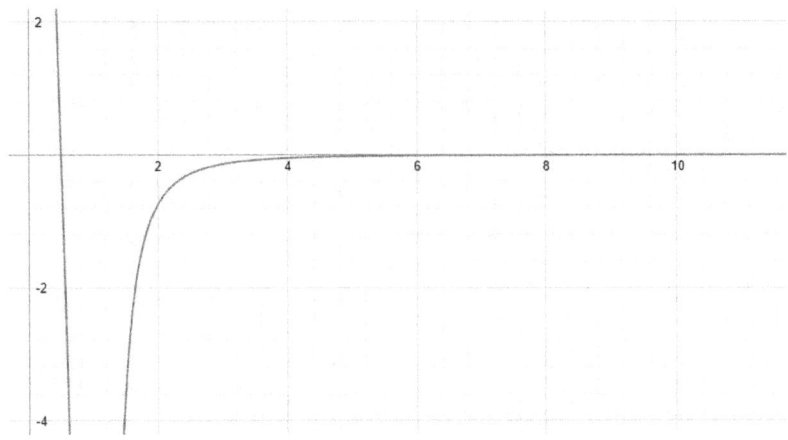

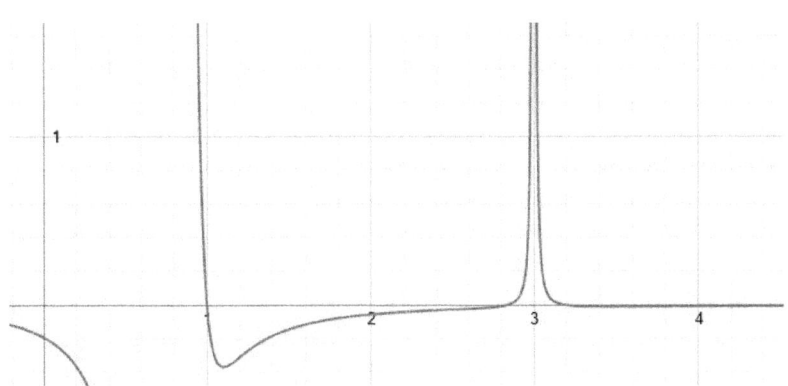

Instead, the various locations of electrons depend on a combination of other positions.

In a 008-O Oxygen, there are:

- Eight protons in a nucleus

- Two electrons sitting at north-south in shell 1 (formerly called 1s2, but called 1m2 to relate to the magnetic poles)
- Two additional electron sitting at north-south in Shell 2 (2m2)
- Four additional electron sitting at 104.5 degrees off the north-south in Shell 2 (2c4)

When you consider bonding, this is outside the $(ZN)* V_{R_{ES}}$, so we focus on charge. Remember that if you set a bond at $(ZN)* V_{R_{ES}}$, then you will have 3 electrons close to the new structure. That electron-electron repulsion is not the low energy spot.

For simplicity, the calculations are:

- Six (6) net positive (+) charges (six protons in the nucleus less two (2) electrons in Shell 1)
- Three (3) negative (-) electron charges located at the same angle off to the side

Near the electron shell, the power of the electron repulsion is greater, so no bonding. However, as you move away, there

#	Dist.	Range 1 1/distance-squared	Force	#	Dist.	Range 2 1/distance-squared	Force	Net Force
4								
4	1.00	1	-4.0000	-3	0.60	2.777778	8.333333	4.333333
4	2.00	0.2500	-1.0000	-3	1.60	0.390625	1.171875	0.171875
4	3.00	0.1111	-0.4444	-3	2.60	0.147929	0.443787	-0.00066
4	4.00	0.0625	-0.2500	-3	3.60	0.077160	0.231481	-0.01852
4	4.25	0.0553	-0.2214	-3	3.85	0.067465	0.202395	-0.01906
4	4.50	0.0493	-0.1975	-3	4.10	0.059488	0.178465	-0.01907
4	4.75	0.0443	-0.1773	-3	4.35	0.052847	0.158541	-0.01874
4	5.00	0.0400	-0.1600	-3	4.60	0.047259	0.141777	-0.01822

Now, this is a simplified example, but it provides the template for the calculation of the bonding distance, and thereby the bonding strength for any atom to another.

The charge force is attractive just about everywhere once you get a certain distance from the electron shell. However, since the force is decreasing 1/distance-squared, then there is a distance of the most attraction, which in the above example is about 4.25 to 4.50 x the electron shell distance or 3.25 to 3.50 as a bond length.

Further, there is a location where the existing electrons are most out of the way, so the bond is strongest in one particular angle versus the stable shell configuration.

Graph Needs Electrons Offset to Avoid Discontinuity (the Deflection of Directly Hitting Electrons in the Shell)

I started this chapter with that graph with the leap off the page. That leaps is only if the electron is directly in line with the bonding atom.

In real life, the atoms go bounce off each other often. However, if they get between the electrons in the shell there is a location where the bonding can occur.

That bonding force is very directional. Sometimes it attracts; sometimes it repels.

- When there is a relatively open path from the distant atom to the nucleus, then bonding happen. From that direction, the shell-vs-proton net force shows as **attractive**.

Side View

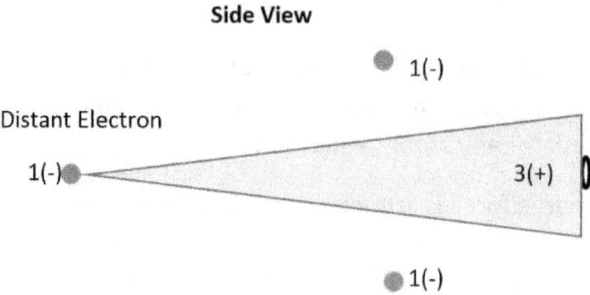

From this direction, the three (3) protons in the nucleus pull more than the two (2) electrons pushes sitting to the side of the nucleus. So, the shell-vs-proton net force **attracts**. *(By the way, the extra electron is on the far side, so please skip the complaint that I made an atom that does not have protons equal to neutrons.)*

- When there is an electron in the way, then the bond does not happen. From that direction, the shell-vs-proton net force shows as **repulsive**. The electron-shell is a barrier.

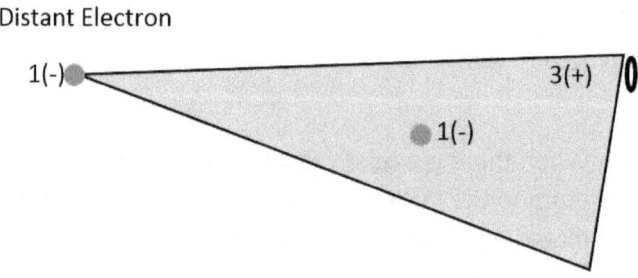

If the distant electron comes from a different direction, then the electron gets in the way, and the electron (-) to electron (-) repulsion overwhelms the attraction form the nucleus.

Remember that 1/distance-squared makes the closer items more powerful. As a result, at some point the distant electron gets repulsed by the closer, 1/distance-squared electron in the way.

As any atom gets closer, electrons first, then that ratio of the distance from electron shell versus the distance from the nucleus gets bigger. With 1-distance/squared, the force balance changes.

Distant Item Location	Nucleus Proton Distance	Electron Distance
3.0x Shell	3 – 0 = 3	3 – 1 = 2
2.0x Shell	2 – 0 = 2	2 – 1 = 1

Using those distances, the two charge calculations change from a net attraction at a distance (far away, like gravity!) to a net-repulsion near to the electron-shell.

This is the above example with one (1) electrons in the way of three (3) protons. You can see the electron create the 'shell' as anything getting close become repelled by the 1/distance-squared electron. The nucleus contains three (3) attractive to the one (1) electron repulsion.

Distant Item Location	Proton Pull	Electron Repel	Net
3.0x Shell	$3/(3^2)=0.333$	$1/(2^2)=-0.250$	+0.083 attracts
2.0x Shell	$3/(2^2)=0.750$	$1/(1^2)=-1.000$	-0.250 repels

Note that the close repulsion is bigger (-0.250) than the far away attraction (+0.083). If it comes any closer, this gets exponentially more repulsive. The attraction as a distance goes on forever. This net-attraction-from-a-distance, atom-gravity, even starts working at a distance only a couple times beyond the electron shell!

As the distant atom gets closer, if an electron is in the way, then it 'bounces' off the 'shell' of the atom. A set of electrons really is an invisible barrier, a 'shell'. In most directions, other atoms bounce off this invisible repulsion distance instead of bonding. Bonding is actually a very special event.

Also, remember that in everything but the simplest atoms (001-H Hydrogen and such), there will be multiple electrons in the way, and that the multiple shells are barriers (1/distance-squared).

Side View – Many Electrons Blocking Bonding Path

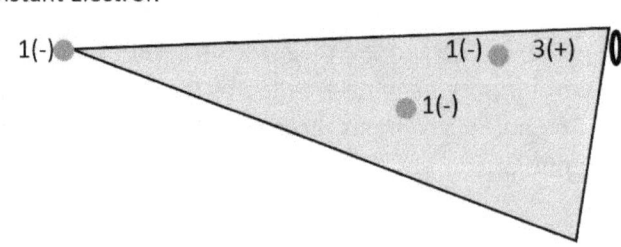

Directional Attractions and Repulsions

Therefore, the bonding attraction has this strength curve, both attractive and repulsive, around each atom.

Electron-Shell-vs-Proton Net Force By Angle

By Direction, Shell-Vs-Proton Net Force is Both + and –

The net-force is sometime positive, sometimes negative. Of course, the full picture is around the 3D sphere, but hopefully the 2D slice above gives the idea. Bonding force is all over the place – positive and negative.

If you take these positive and negatives, the spherical approach to atoms and molecular bonding, you do not get this 'discontinuity' except if exactly approaching through an electron in the electron shell.

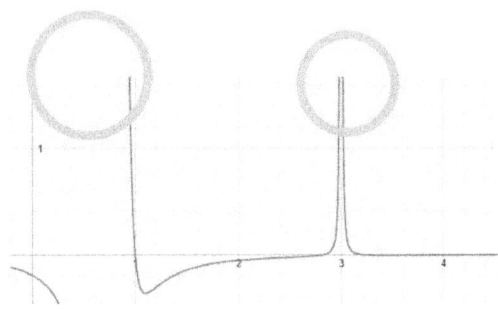

Instead you get a net repulsion at the electron shell, but an overall attraction at 3-5 times that distance.

This is the curve of force for approaching electrons of another atom trying to create a molecular bond. It is the force the existing electrons in the shell.

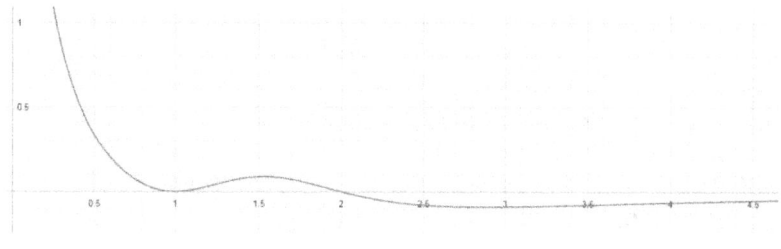

$$■ - \frac{1}{(x-1)^2+1}$$
$$+ \frac{1}{(x-1)^3+1}$$

This is electrons that are to the side of the approach. That means they has a distance plus the extra of the standard electron separation (f + 1).

Wow! Looks at what happen. We have three amazing points.

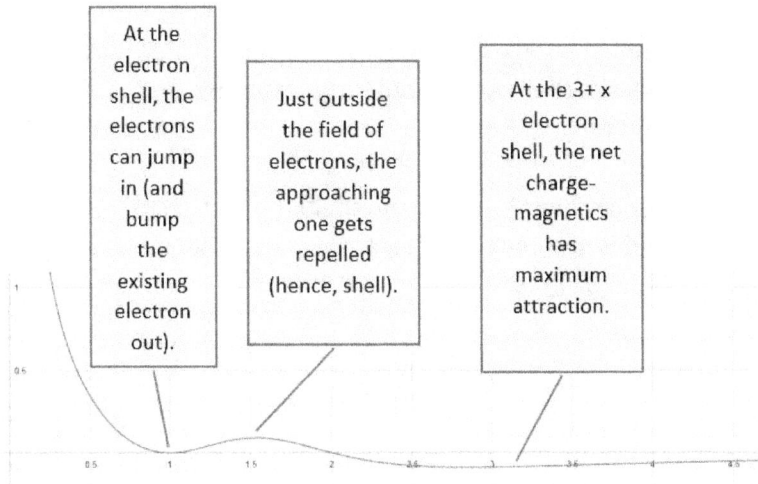

Now, the actual calculation is a little more complex. It has the approaching atom as the electron/nucleus set with opposite charges. That accentuates the three points, but does not change the general conclusion.

Bonding energy is a) the maximum attractive charge-magnetics force to the nucleus (proton attraction) given that other electrons in the shell has the repulsion but express that repulsion at an angle which makes them slightly weaker.

The approaching-electron graph shows various interaction distances:

- If the new molecule penetrates the electron shell, it is repulsive (+ on the graph).

- At zero, the net force is zero, and sometimes that means the new electron can bump an existing electron out.

- For a range (1-2x electron shell) just outside the electron shell, the force is repulsive. An electron-shell really does act as a shell keeping other electrons and atoms away for that range.

- For a range (3-4x electron shell) the attraction force exists and is maximum. That is the distance that molecular bonding likes to occur.

Connection of Charge-Magnetics to Mass

Back to main subject, the full magnetic field causing electron shells is in direct ratio the sum of protons and neutrons in the nucleus. Therefore, it become a linear solution based upon nucleus particles (=mass).

So, that relationship between the electron distances is consistent (with the adjustments). The nucleus has XX particles and magnetic strength; therefore, the mass has the same ratio to those XX particles. We find that mass is directly related to the number of nucleus particles.

Most importantly, we find that the average distance of electrons in all shells is directly related to the XX nucleus particles.

That means that in Newton's Gravity, and Einstein's Energy we can substitute for *m* with the integral:

$$m = \int_0^{Z+N} \left(\frac{M(\#_{ZN})}{\frac{8}{3} R_{ES}^3} \right)$$

For the entire proof of that substitution, read the other book in this series. <u>Gravity is Just . . . That Electrons are a Little Closer</u>. It takes an entire book to work through the details.

Postulate #8: The Fundamental Charge-Magnetics Function is a double-function operating from and delivering results to both charge and magnetism as two output values in coordination.

Today, forces get analyzed as one force or the other, charge or magnetism. The confusion in current textbooks flows from the lack of providing both outputs, and their structure based upon the distance range.

Or worst, the input or results get restated as some unknown name. We have separate formulas that include 'spin', 'color', or 'magnetic moment' for the cause or 'strong interaction', 'weak force', 'gravity' for the result.

Instead, consider a formula that has only the two factors, and two results, yet that is two – a*t the same time*. Flipping between charge and magnetism using this basic model, you get one-double function with the charge function and result and the magnetics function and result which are inextricably linked going into the formula, and coming out of the formula.

This double-function is important to find and understand missing Newton reactions. The full action and reaction will get found.

Some previous solutions have these multiple-output structure. It uses matrix mathematics. Therefore, people with current experience should understand this multiple output structure. However, this solution is a matrix with only 2 outputs, and those two are related. Therefore, the solutions are ultimately simpler than the heavy challenges of current NxN matrix models with more criteria where the matrix has 3, 7, or even 18 functions.

The matrix of 2x2 is much simpler:

$$= \begin{pmatrix} 1 & 0 \\ -1 & 1 \end{pmatrix}$$

Than a matrix which covers charge, 'spin' and 'color' and creates formulas that look like:

$$= \begin{pmatrix} 1 & 0 & 0 \\ 0 & \cos Rt & -\sin Rt \\ 0 & \sin Rt & \cos Rt \end{pmatrix} \begin{pmatrix} U(0) \\ V(0) \\ W(0) \end{pmatrix}$$

This postulate is important because many useful calculation use one side, 1-way charge, and the balancing aspects, the magnetic, and not observed or not integrated in current textbook calculations. That integration and use of those Newtonian opposites, sometimes invisible, are one of the key points of understanding.

Newton's Gravity Formula Revised

Remember that we introduced that the 'mass' is the integral sum of the magnetics forces on the nucleus particles $M(\#_{ZN})$, given that separation distances of the radius of the electron shell (R_{ES}).

$$G\frac{m_1 m_2}{d^2} = \frac{\int_0^{Z+N}\left(\frac{M(\#_{ZN})}{\frac{8}{3}R_{ES}^3}\right)}{d^2}$$

We can eliminate the $\frac{1}{d^2}$:

$$Gm_1 m_2 = \int_0^{Z+N}\left(\frac{M(\#_{ZN})}{\frac{8}{3}R_{ES}^3}\right)\int_0^{Z+N}\left(\frac{M(\#_{ZN})}{\frac{8}{3}R_{ES}^3}\right)$$

Given that both sides (for the calculation of 001-H to another 001-H at 1 meter) are the same Z+N, then we can equalize and square root to get:

$$\sqrt{G}m_1 = \int_0^{Z+N}(\frac{M(\#_{ZN})}{\frac{8}{3}R_{ES}^3})$$

Then solve for m:

$$m = \frac{\int_0^{Z+N} \left(\frac{M(\#_{ZN})}{\frac{8}{3} R_{ES}^3} \right)}{\sqrt{G}}$$

NOTE: The square root of G is really a fudging factor for the radius of the electrons shell, plus the extra integral 2 and/or polynomial 2 or 3. I will get that works out and add those steps here when ready. Please understand that the square root of G gets eliminated.

When Newton created that calculation of gravity, he used a factor called 'mass' which solved all the formulas. However, 'mass' even centuries later, scientist never figured the connection how the gravity force related to the core particles: protons, electrons, and neutrons. In fact, scientists keep searching for some 'god' elementary particle that has this gravity force that only works at a distance.

Yet, the best connection is the number of particles in a nucleus.

- At one particle, 001-H Hydrogen, the formula works. This is very helpful since starts consist mostly of 001-H Hydrogen molecules

- For the large molecules, the comparison has a slight different called 'mass deficit'

Mass can get replaced in the Gravity formula by the calculation of the nucleus-particle-magnetics of the volume that field encloses (Gauss).

$$Energy = \int_0^{Z+N} M(\#_{ZN})$$

The reason for the integral is that M() magnetics is oriented, and some of the nucleus particles (ZN) physically orient to create partial cancelations. Often, this would get presented as:

$$Energy = M(Z + N)$$

When done this way, Bohr was able to calculate for the simplified Z = 1 and N = 0, the Energy constant in terms of Planck and the Speed of Light.

This Energy is spread over a volume of the electron shell:

$$Volume = \frac{4}{3}\pi R_{ES}^3$$

That gives a field strength formula of:

$$Strength = K - \left(\frac{\int_0^{Z+N} M(\#_{ZN})}{\frac{4}{3}\pi R_{ES}^3 x^3}\right)$$

Remember that Bohr ended with an extra factor a_0 to make the calculation work. That is the fudge-factor for $Volume = \frac{4}{3}\pi R_{ES}^3$ and the fundamental relationship of Charge and Magnetism.

And, the integral of that strength, in a particular direction, to a particular distant object, is the Gravity Force.

$$Force = \int_{x=0}^{x=d} \left(\frac{\int_0^{Z+N} M(\#_{ZN})}{\frac{4}{3} R_{ES}^3 x^2}\right)$$

Einstein's Energy Formula Revised

Remember that we introduced a richer detail of 'mass' that have two elements, a) the magnetic force of the nucleus particles, and b) the differential caused by the radius of the electron shell.

$$m = \frac{\int_0^{Z+N} \left(\dfrac{M(\#_{ZN})}{\frac{8}{3} R_{ES}^3} \right)}{\sqrt{G}}$$

From the previous, we can now run full circle from Bohr/Planck to Einstein.

Einstein states that he has computed total Energy, but instead he has computed the fundamental Energy over a particular volume. That volume is very important because it is the natural distance of the Charge-Magnetics relationship. It the natural point of balance towards which nature will fall back.

Einstein should get clarified to say.

$$\frac{Energy}{Steady\ State\ Volume} = m(c^2)$$

And that steady-state only gets achieve at the speed of light. That is the time that it takes for charge-magnetics to balance at the Bohr radius.

Therefore, Einstein missed that this is not the actual total energy. That energy is the M() magnetic charge of the particles.

By substituting into the

$$\frac{Energy}{Steady\ State\ Volume} = m(c^2) = \int_0^{Z+N} \left(\frac{M(\#_{ZN})}{\frac{8}{3}R_{ES}^3}\right)(c^2)$$

But again, that Volume ($\frac{4}{3}R_{ES}^3$) is really misleading. We can multiply both sides of the equation to get the total energy (without the Steady State limitation).

$$Total\ Energy = m(c^2) = \int_0^{Z+N} \left(\frac{M(\#_{ZN})}{2}\right)(c^2)$$

That is great. The total energy is the Total Energy of the Magnetism.[xxii] The magnetic charge is a limiting factor and the total energy of a system. The magnetic charge M() can only get pushed out at the speed of light. Now, Einstein makes perfect sense.

Why does Einstein say that space and time warp?

Einstein's solutions works. If you make time or space warp, you get the same solution as if you change R_{ES}. The R_{ES} is a constant in everyday situations; R_{ES} changes when a) the system is moving at a % of the speed of light because the electron on the back of the rotation slow down stretching the R_{ES}; and b) R_{ES} reduces for bonding electrons because their magnetic field is caught between the combined nucleus of the multi-atom molecule system. The bonding electrons contribute to the charge separation of the electrons, but do not contribute to the magnetic function. Again, this an example of the two-force matrix of charge-magnetics.

Fundamental Question: Why does 'mass' change at speeds near the speed of life?

"Mass is found to increase with velocity, but appreciable increases require velocities near that of light."[xxiii]

Another way that scientists have observed that 'mass' changes is for atoms moving at a material percentage of the speed of light. However, when you substitute in the deeper calculation of observed 'mass' from the above, one factor, R_{ES}, makes that 'mass' change understandable, and the other, $(Z+N)$, does not.

When moving at the speed of light, the electrons orbiting have a challenge on the back side, they have speed limit so the electrons are slow to catch the moving nucleus. As a result, the electrons shell, and thereby, R_{ES}, gets 'stretched'. They make it around, but because of the speed, they have a longer path, and thereby larger R_{ES}. The $(Z+N)$ does not change as observed through multiple experiments of before, during and after.

There are again multiple issues in this calculation:

- Angular momentum, the speed of the electrons, as increasing make the same field strength have a larger radius

- Magnetism created by the particle or whole-system movement may increase the attraction

This is great because we can isolate the change to one half, and the other half, the number of nucleus particles can remain constant. The number of nucleus particles, **N**, does not change. At the end of the interactions causing space-time calculations are a physical known, we still have the same element as each atom. We know the ending atom is exactly the same as the initial atom in number of protons, neutrons, and electrons.

Fundamental Question: What Explains why Mass Changes in Bonding?

Wikipedia says:

"Classically a bound system is at a lower energy level than its unbound constituents, and its mass must be less than the total mass of its unbound constituents."[xxiv]

What changes is R_{ES}, not (ZN) (the number of nucleus particles).

When you add another atom and share the electron between them, that structure does not have the protons all in the middle. That changes the R_{ES} calculation. The space between new combo-nucleus is a dead zone, and that is exactly where the bonding electrons sit.

The total # of N nucleus particles does not change, but number of those particles that contribute and how they contribute to the gravity calculation do change. The bonded electrons stop part of their contributing to the gravity formula.

To me, M(NP) is the underlying source of mass, the energy element, not $$m = \int_0^{Z+N} \left(\frac{M(\#_{ZN})}{\frac{8}{3}R_{ES}^3}\right)$$ which is the 'observed mass'. R_{ES} is a reducer as that volume over which that mass energy got spread at atomic steady state. Only because R_{ES} is a direct relationship to (NP) (but not direct to charge), then the 'observed mass' and 'mass' solve all equations equally as well. Of course, Bohr had to add permeability for that translation. Newton had to the Gravity constant for its translation. Each solution has some magic, but constant factor to implement the ratio of Magnetic Force, M(), with the 'observed mass' $$m = \int_0^{Z+N} \left(\frac{M(\#_{ZN})}{\frac{8}{3}R_{ES}^3}\right)$$.

The extra nucleus in a different centering point changes the average distance of the set of electrons. This factor is smaller, but offsets the bonding lack of contribution.

Changing the average distances creates a slight change in the Charge/distance-square electron-vs-nucleus differential. That changes the force projected to other atoms – the 'mass.'

As such, the differential is slightly different, and we can measure the change in 'mass' which is a change in the average electron distance of the combined molecule versus the atoms operating autonomously.

Bonded Electrons do not contribute to the external Magnetics (Atom-gravity) 'mass' observed

A picture explains. In an atom, the electrons must be all pushed out relative to the nucleus. That gives a consistent ratio which measures as a steady mass. However, in a molecule, that is bonded atoms, the nucleus sits in different places. As a result, those electrons that are between the nucleus1 and nucleus2 do not contribute to gravity. In fact, they have the reversing effect, the nucleus are outside those electrons so that reduces the atom-gravity.

Bonding Electron Not Contributing to Gravity / Mass

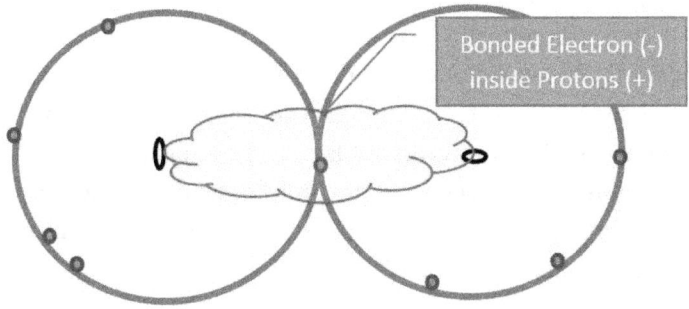

The electron in the bonding position does not contribute to charge-differential (atom-gravity 'mass'). In fact, it decreases the calculation overall. Instead of electrons outside protons consistently, this event is electrons inside protons.

Bonding – As If Electron Removed From Mass

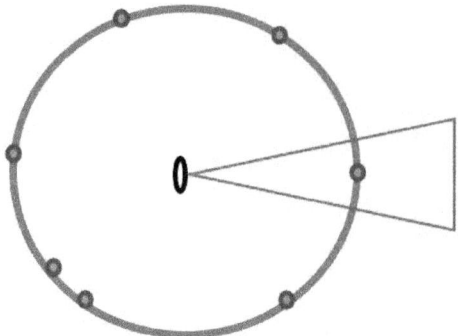

But there are 'reflections' that mitigate that. One great set of mathematics described something similar as the Fresnel zone. In my real job, I have built cellular telecom systems and more specifically microwave towers to connect those towers. We needed to check Fresnel for each tower. Microwaves fail even if the line of sight is clear because of the other 'Fresnel zones.' Between two towers in a narrow window (like the electron-bond cloud), in increments of 1/n wavelengths, that Fresnel gets calculated, if there are any objects in that red cloud, the signal gets messed up. That calculation can get messy, but generally, all you really need to know is do not let trees even in the 'curved space' between microwave towers. A tree, wall or parked truck even not directly between can creates reflections that change the base microwave operation.

The slightly different logic in the details applies to atom-gravity 'mass', but it has similar complexity. While the electron does not contribute to charge-differential (atom-gravity 'mass'), there are 'reflections' that mitigate that.

One would expect that the 'mass' lost in bonding would get reduced by that reversal of:

Electrons bonding / Total # of Electrons

However, experimental evidence shows the change is not that dramatic.

Decrease offset in that Electron still contributes to push out other electrons

It is one step more complex. While that electron does not create a charge-differential, it does contribute to keep the other electrons out in their orbit which means that the average radius does not decrease.

If you take out that electron, then the other electrons would have been closer in. That electron, even in the bond, still contributes to the electron-shell radius (distance for 1/distance-squared).

Without that bonding electron, the other electrons would have position closer together. Without that electron (red triangle) the other electrons (purple) would not have the electron-electron repulsion (orange arrow) from that missing direction. As a result, the bonding electron still does some contribution to the electron-shell radius.

Bond Electrons Do Add Part of Mass that Pushes Other Electrons Outside Proton-Proton Vortex

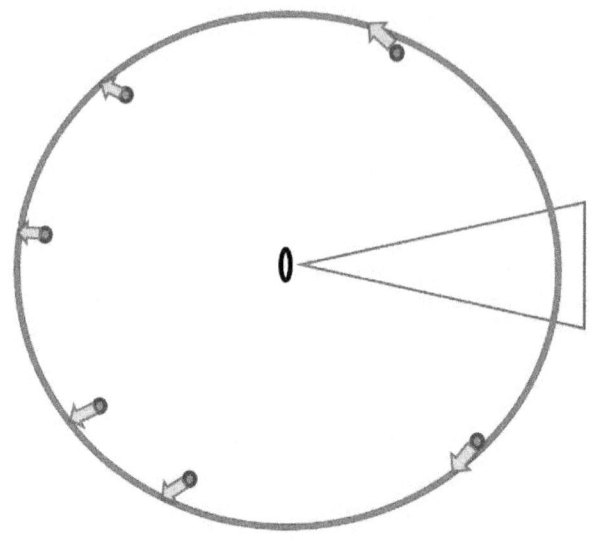

The electrons-in-bonds impact on shell distance can get describe as:

	Electron-Electron Repulsion	Electron-Electron Repulsion	Change to atom-gravity
In Atom	100%	100%	
When Bonded	100%	(n-1)/n	Pro rata

Decrease offset by increases in distance to electron in other atom in the bond.

In addition, the bond distance also now goes in part to both nucleus locations. That makes distance (and thereby atom-gravity) increase.

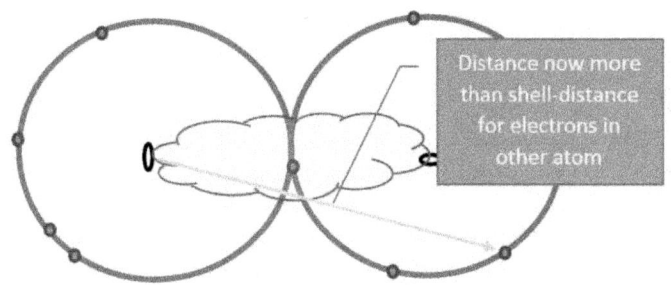

Must also realize location is not fixed, but a quantum path

Further, each electron floats, so, in its quantum path, that electrons actually is outside that zone some percentage of the time and does contribute in those periods.

Calculations done with a perfect circular orbit or exact electron location yield a result up to a factor of 0.6 too small. However, in quantum-path, that electron predicts to be closer and farther away. However, again the 1-distance-squared makes the times when quantum-walk closer have huge impact.

Conclusion: Too much to calculate here

So, I will report in a few years and update this book when all those elements get calculated. I have not done those.

However, in generally, the atom-gravity ('mass') of the combined molecule must go down when compared to the individual atoms. I expect the mass reduction is a Frenzel zone calculation. Those electrons shared contribute less than standard to the charge-differential calculation that is the basic of atom-gravity ('mass').

Subatomic Particle Forces

In a range smaller than the nucleus particle size, many observations have occurred.

First, at this level, magnetism totally rules. Even if a charge got closer, the magnetic force would still be stronger.

$$F = k \frac{Q_1 Q_2}{d^2}$$

Factor	Charge-Force Estimation
Charge-Force Constant (Coulomb)	$k_e = 10^{10}$ [8.99x10^9 m^3 kg^2/ (s^2)]
Charge Proton	$q_1 = 10^{-15}$ [0.9×10^{-15}]
Charge Proton	$q_2 = 10^{-15}$ [0.9×10^{-15}]
Distance	d=3.0 x 10^{-16} m or 10^{-16} m
Exponent shortcut	+k+Q+Q-d-d
Short-cut calculation	10-16-16-(-10)-(-10)= 10-16-16+15+15= 10-32+30=8 N (Newtons)
Charge-Force Repulsion	10^{+8} m^1 / (s^2) (Newtons)

$$F = M_A \frac{(NP_1)(NP_2)}{d^3}$$

Factors	Magnetic-Force Estimation
Charge-Force Constant (Coulomb)	$M_A = 10^{-38}$ m³ kg²/ (s²)]
Charge 001-H Hydrogen atom	n=1 or 10^0
Charge 001-H Hydrogen atom	n=1 or 10^0
Distance	d=0.9 x 10^{-15} m or 10^{-15} m
Exponent shortcut	+k+Q+Q-d-d-d
Short-cut calculation	-38+0+0-(-15)-(-15)-(-15)= -38+0+0+15+15+15= -38+45=+7 N (Newtons)
Magnetic-Force Attraction	10^{+7} m¹ / (s²) **(Newtons)**

The charge force at a nucleus distance would be:

$$F = k \frac{Q_1 Q_2}{d^2}$$

Factor	Charge-Force Estimation
Charge-Force Constant (Coulomb)	$k_e = 10^{10}$ [8.99×10^9 m^3 kg^2/ (s^2)]
Charge 001-H Hydrogen atom = 1 proton	$q_1 = 10^{-19}$ [1.67×10^{-19}]
Charge 001-H Hydrogen atom = 1 proton	$q_2 = 10^{-19}$ [1.67×10^{-19}]
Distance	d=0.86 x 10^{-15} m or 10^{-15} m
Exponent shortcut	+k+Q+Q-d-d
Short-cut calculation	10-19-19-(-15)-(-15)= 10-19-19+15+15= 10-38+30=+2 N *(Newtons)*
Charge-Force Repulsion	10^{+2} m^1 / (s^2) **(Newtons)**

Yet that has to be less than the magnetic repulsion at that same distance – which it is as shown above.

So, magnetic-force is larger compared to charge-force at the same nucleus particle distance. That is, if the particles are separated by the distance of a particle.

However, the real comparison is that charge-force, if proton-near-proton is much less. If no separating particle, the protons get much closer as shown below. The proton-proton repulsion is huge, even versus the magnetic attraction.

Question:

Nucleus has not glue, no gluon, so a sideways force will break the nucleus structure. The particles float in paces. Of course, it has to be stronger that magnetics which at that distance is very, very powerful. That break is basis of nuclear decay, the structure gets knocked, and the proton collide.

What is a Neutron?

The evidence shows that a neutron has a statistical profile that also has charge building. It builds a positive charge, then has a shell that balances it back to zero at the edges. To me, that means that the edge is negatively charged.

However, that model is statistical. It looks at distance, but not direction.

The conjecture based upon the postulates is that a neutron is combination of a neutron and a proton; or a neutron, a proton, and a photon.

Based upon the postulates so far, we can explore geometry. The magnetic strength is 10^{+7} Newtons but in what direction. What is the strength in every direction? Which direction might create both a place for electrons and/or protons to bind? And leave for binding with another proton?

The natural answer is north-south. At the poles, the magnetic strength is small, so that would be the natural path for electrons

Think about three possibilities. First the electrons sit at ends:

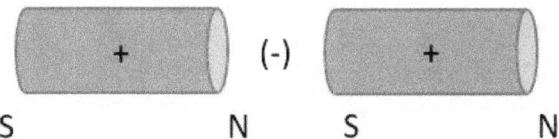

Second, the protons rotate creating magnetics sit at ends:

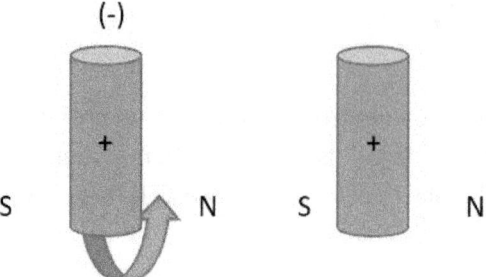

Third, potentially, there is a four particle structure at this boundary of touching where the combination is a 'neutron'

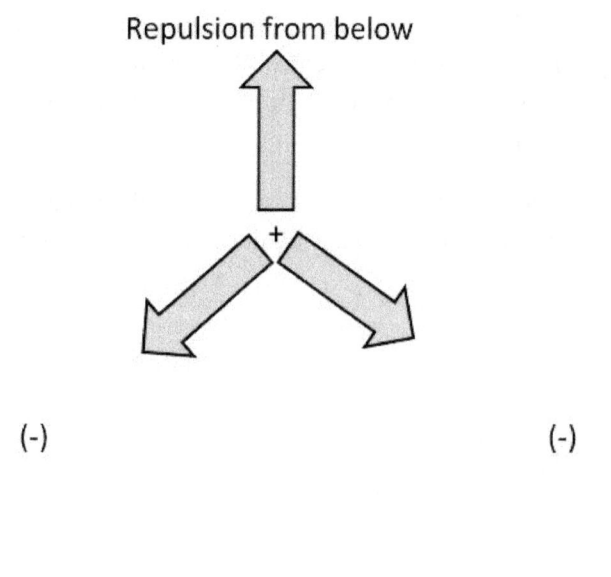

With four particles, one particle gets a triangle of vector force such that the opposing ones, so that for a particle electron:

      1x   Proton-Proton

      2x   Proton-Electron, so each move ½ in direction X

But as vectors, the size must be at angles, so by Pythagorean's Theorem:

      Cos(Angle) = ½ = 0.500 which is 60 degrees

Further, this structure will be magnetic perpendicular the center of the ring, and thereby that magnetics

The distance of the protons is 1 (½ plus ½)

Because this is 60 degrees, the sin(60) = $\frac{\sqrt{3}}{2}$ which is $b^2 = 1^2 - \left(\frac{1}{2}\right)^2 = 1 - \frac{1}{4} = 3/4$. The distance of the electron separation is $\sqrt{3}\,(\frac{\sqrt{3}}{2}$ plus $\frac{\sqrt{3}}{2})$.

Therefore, in this structure, the electrons are a little further apart, but the overall structure is stable.

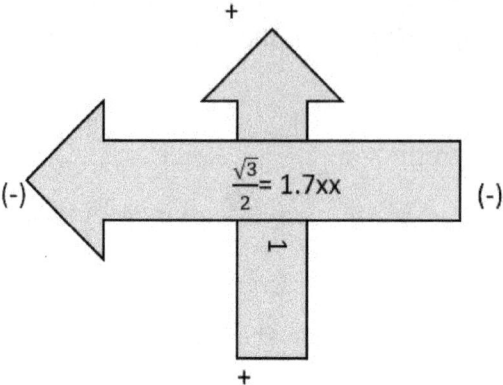

This corresponds to experimental evidence that the electrons sit at the outside of the rotating structure. That the profile of a neutron is a more positive core, changing to zero only at the exterior.[xxv]

The sub nucleus items might be the alternatives of:

    Protons with electrons at one pole

    Protons with electrons at both poles

    Electron proton pairs

Again, the solution is driven by magnetics at these distances.

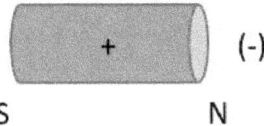

S　　　　　N

S　　　　　N

The observations that as you pull apart certain sub nucleus particles, that it creates two seems the basic nature of magnets. If you have a magnet and cut it in half, you have two magnets.

Fundamental Challenge: All Work in Subatomic Particles Done in Charge (eV), but Not Enough Done on Magnetics?

Based upon the net charge-magnetics described, why is all current research using eV charge as the measure of subatomic particles

Atomic Mass Used in Many Force Calculations

In many calculations, they have used a measure of mass, but what is mass?

Mass is consistent.

Mass is related to the atomic mass which is the number of nucleus particles.

However, what we observe as mass is a net force.

Because it is for distance object, charge is more powerful. As a result, it is the far end of the range. That means that only charge works at that distance.

However, we do have two particles with charge, the protons in the nucleus and the electrons in the shell.

This the most complex of force formulas. It is a charge difference, but that difference gets driven by the magnetic force balance.

It is the only complex formula that has two solutions. One is a calculation based upon charge based upon the electron shell. However, that electron shell is a balancing point of charge versus magnetics; therefore, the nucleus particles, which is a magnetic source substitutes into the same place.

That means you can use either calculation depending on the circumstances.

What is the magnetic field of a nucleus particle/nucleon (proton, neutron)?

The field gets calculated based upon the strength less the distance to the two poles. This applies to the magnetic fields of the nucleus particles/nucleon (protons or neutrons).

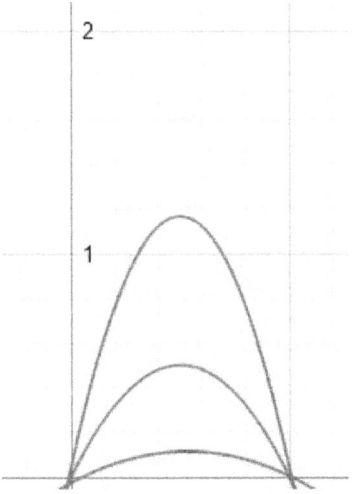

This graph looks like the results for iron filings and an iron magnet.

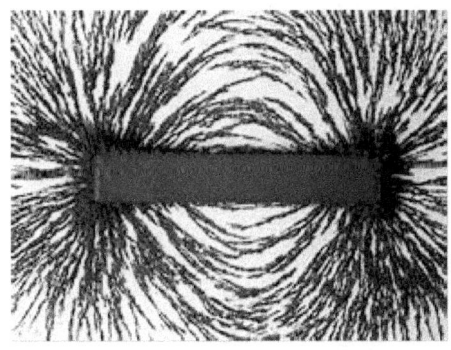

The magnetic field is actually strongest at the corners. That is where the differential is greatest; the space moves from magnetic to non-magnetic so the differential is the observed strength.

Is that is what is happening at the radius of a nucleon (proton or neutron)?

What is a Neutron?

A neutron is a magnetic proton with an electron sitting at the exact middle of the proton magnetic direction.

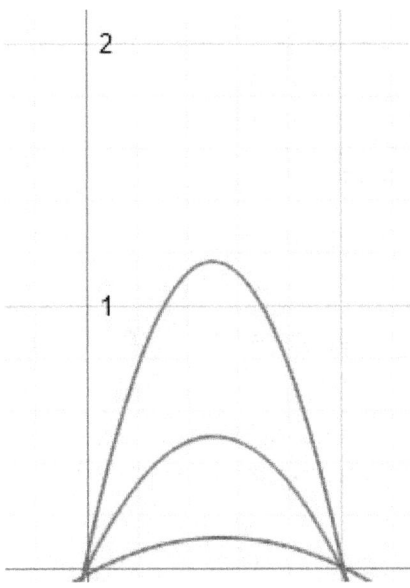

Question: Why is it important the Magnetic Force Overwhelms Charge Force at the Sub Nucleon Distance Range, but Multiple Charges are Not Observed?

One of the areas for further exploration comes from the revelation that charge does not matter below 10^{-16} m. Everything gets driven by magnetics as the dominant force.

That means that protons could stick to protons.

Yet, we have centuries of experiments that refute that without the intermediate neutrons.

I revised Pauli exclusion for electron shells to a geometry alternative solution in <u>Scrunched Cube</u>.

The graph of a bar magnet is not continuous. In fact, the magnetic strength comes from the places of most 'differential' which is the corners. Here is a picture of that:

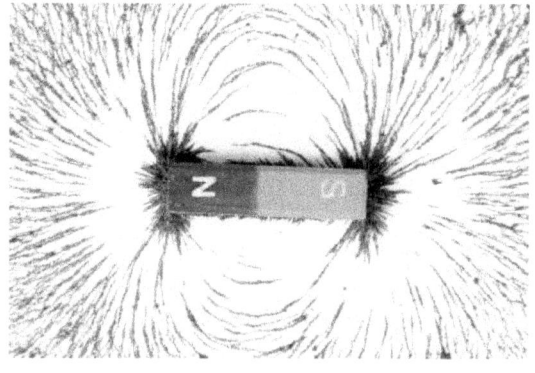

The bar magnet is actually stronger at the poles, not exactly like the strength graphs show. That means that electron binding to protons could occur off the middle so thereby creating a neutron. That charge attraction creates a strength larger than the magnetics at those distances.

The reason is that is the highest differential, to open space, versus must magnet to magnet. In effect, at the speed of light, the movement of charges, is that nucleus distance.

$E = mc^2$ and the Fundamental Ratio of Electromagnetism

The fundamental ratio relates directly to Einstein's $E = mc^2$

Transforming $E = mc^2$ for the underlying fix of 'mass' into its components means the substitution of $m = \frac{MN_1}{\frac{8}{3}\pi(R_{ES})^3}$.

Let's look at the units of measure, and an interesting thing happens.

The definition of mass that 4π.

$$m = \frac{\int_0^{Z+N} \left(\dfrac{M(\#_{ZN})}{\frac{8}{3}R_{ES}{}^3} \right)}{\sqrt{G}}$$

The definition of the Bohr radius that 4π.

$$a_0 = \frac{4\pi\,\varepsilon_0 \hbar^2}{m_e e^2} = \frac{\hbar}{m_e c \alpha} = R_{ES}$$

The Newton definition of gravity does not have 4π since it is not at over a sphere, but to a distance point from a point:

$$F = \frac{Gm_1 m_2}{d^2}$$

But, the revision does because of the spherical relay:

$$F = \frac{Gm_1 m_2}{d^2} = \frac{\int (M(ZN)_1) \int (M(ZN)_2)}{d^2 (\frac{8}{3}\pi R_{ES}{}^3)}$$

The definition of charge should not have 4π because it assume a point source:

$$F = k\frac{Q_1 Q_2}{d^2}$$

Education Opportunities Based upon Postulates

Resolving and Restating Mass Changes Everything

Feynman, in his famous series of CalTech lectures, said about the basic structure of physics thinking and how current theory require laws to get set aside: "Finally, and most interesting, philosophically we are completely wrong with the approximate law. Our entire picture of the world has to be altered even though the mass changes only by a little bit. This is a very peculiar thing about the philosophy, or the ideas, behind the laws. Even a very small effect sometimes requires profound changes in our ideas."[xxvi]

To me, the core philosophy build from determinism is not wrong. The whole direction that Feynman encapsulated shows the weakness of just certifying any solution that fits the experiment observations. A is A. Magnetism of nucleus particles does not change; the 'cause' of 'mass' does not change. Instead, we missed a relay point at electron shells. The size of the relay point changes so the 'observed mass' changes in very limited ways, but remains generally in a direct relationship to particles. This re-orientation fixes Newton's gravity; this fixes Einstein's energy.

The choice to use 'observed mass' in the Newton Gravity formula leads to Einstein warping space and time to solve a problem in Special Relativity and General Relativity. It leads to Einstein using 'mass' in $E = mc^2$. But variances still occur because 'observed mass' is not constant, particle magnetic energy is. Therefore, scientist invent other dimensions and abstracts that solve the experimental evidence and anomalies. They took the incomplete concept of 'mass' and then added stuff so it might work, and making it impossible complex. Hence, the need for Feynman's long explanations of the complexities of 'fit to experiment' current statistical models.

Reintroducing Geometry Replacing Statistical Solutions

Further, geometry is missing for many solutions in Feynman. For a century, everyone thinks about only statistical solutions because of missing concepts which statistics provided workable solutions. Even Einstein worried about this trend. The story is that Einstein invites Schrodinger, who reasoned out that complex statistical math, for a weekend visit, and the guest, Schrodiner, after that weekend, then writes his famous Schrodinger's Cat thought experiment to demonstrate the absurdity of his own direction of particle physics. Of course, people now think Schrodinger's Cat certifies probability science, instead of mocking it.

A clear course of education will utilize geometry as fundamental which will open the door to a student that is visual, versus only an abstract math wiz. It will make particle physics something that reaches millions versus just a few.

Students cannot and do not become physicist because they must set aside reality to complete the coursework. Yet, they must accept the entire chain of Feynman, and live is some virtual 7-dimensional string-theory world to get their degree. There are hundreds of complexities and extra-dimensional steps that Feynman lectures must get accepted. That is why a majority of people cannot fathom the current path of coursework. The Feynman documentation of all those 'fixes' to match experiments need to get revised, and replaced with cleaner postulates. In reviewing those, I find that the postulates here would revise something on almost every page on those Feynman lectures.

The entire physics and chemistry teaching methods could change towards deterministic fundamentals. Instead, the postulates will create a fundamental movement of the pendulum back towards Newton. Complexities go away that require a physicist or chemist to become an expert in abstract mathematics.

Conclusions and Opportunities

There are more opportunities in the use of these postulates. I have more potential ventures than I have time to follow. Remember that I come from business. I do not work at universities. As much as this is theoretical, the applications and value become real products:

- I am developing the potential from the electron shell postulates for quantum computing are in my field of expertise; you see, I built the first GSM cellular systems in various locations from the former Soviet Union to the Guyana jungles of South America. In part, it is because my professional career included so much on electromagnetic fields that I saw the connections explained here.

- I have a chemical company to whom I consult with potential applications that can reduce energy costs dramatically, and improve chemical plants and processes.

The calculations of other relationships could fill dozens of more PhD proposals for others to pursue. It can change water purification. It can improve carbon utilization.

And, yes, I am a homeschooling father. I think that we can simplify science education, and the opportunities for the next generation are boundless. The current university physics require people with my level of skills or greater: a) in matrix analysis, b) in quantum statistics, and c) in intuitive abstract mathematic. However, I believe that basic high school geometry, and a little vector math, and we will have tens of thousands of chemistry university graduates that will solve problems beyond what I can.

This is bigger than me. Please ask questions . . . please challenge me. I am here to help! I am lucky enough that my success allows me the time to explore. However, I can achieve more when others join together. I welcome the collaboration and competition because the entire system needs reworks based upon these postulates.

I know my biggest fears today is that I forgot a factor 3 or 2 in some the integrals of polynomials with negative exponents in the field strength in the gravity proof, and I have not found anyone else with the brainpower to check every detail of the calculations in various postulates and books. Please help!

Big hugs. Let's go change the world.

Arno

Arno Vigen Science Postulates:

For more than a century, the pendulum of physical sciences moved away from Newton and the concrete, physical world. The four Arno Vigen postulates below move the pendulum one step back towards center. A is A. Physical reasons are better describers of physical science, when those physical factors are discovered.

While other solutions get the correct answer in brilliant, amazing, creative formulas, the deep answer becomes simple and real. It becomes something that we can teach to every student, without LaGrange, Hamiltonians, and Gaussian differentials -- without, or better said resolving into the basics, the invented 7, 9, 11 or 18 dimensions of the latest version of string-theory.

The four postulates go back to the basics:

- Three physical dimensions (length, width, height or their spherical equivalents)

- Time

- Electromagnetic fields*

- Known particles – protons, electrons, and neutrons

There are two separate currently until someone like me discovers the direct interconnection.

Each of the postulates takes current, complex calculations, and gives them a clean path using only the above basics.

#1 Electrons repel magnets – *both poles*

- Resolves what force makes the electrons stay in a shell
- Resolves spin number in various subatomic particles
- Resolves color factor in various subatomic particles

It simplifies to the below two graphs that explain the charge force versus the magnetic force. Both decrease, but for a particular point, the charge force is always stronger (chart 1), but in chains the charge is not a point, but a chain which means it does not decrease until the end of chain – keeping it strong enough to stay linked even if protons repel.

It creates a graph of the Newtonian action-reaction forces of charge ($\frac{1}{d^2}$) and minimum-north-south direction magnetics ($\frac{1}{d^3}$).

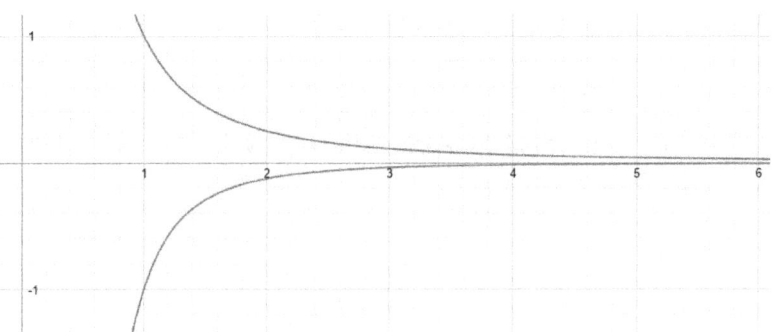

Which becomes at combined charge-magnetics net-force:

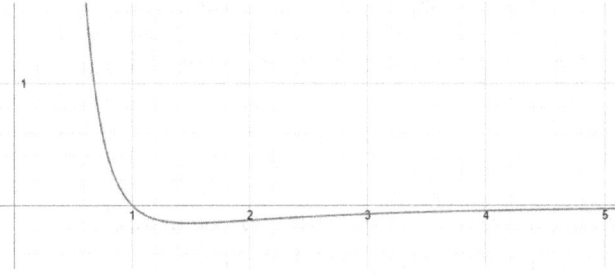

Charge is stronger than magnetism outside the electron shell. Magnetics is stronger than charge inside the electron shell. The average distance of an electron shells is the balancing point.

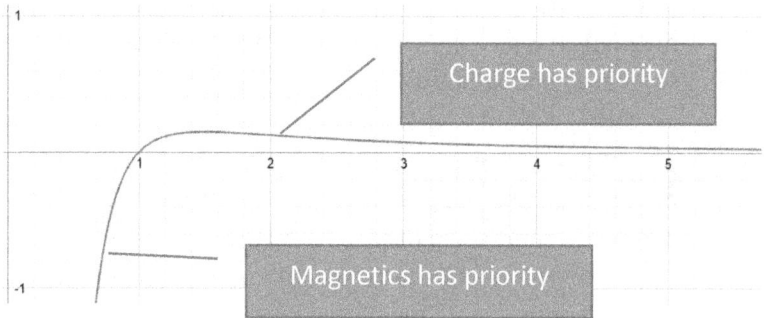

#2 - Every nucleus holds together via a chain/ring-particle-magnet organized as proton-neutron-proton-and so on:

- Resolves what holds the nucleus together (strong force) gets based upon when in a chain connected along an oriented magnetic field, that, at the nucleus distance, magnetism is stronger than charge if the charges are separated enough by an intermediate neutron.

- Educational nucleus-plus-chemistry set patent 15256865 pending relating the magnetic field of the particles to the overall magnetic field of the chain or ring nucleus structure with north-south perpendicular to the ring as shown below.

Figure 12

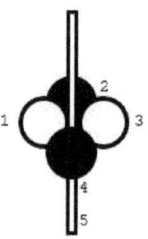

Charge is stronger than magnetism outside the electron shell. Magnetics is stronger than charge inside the electron shell. The average distance of an electron shells is the balancing point.

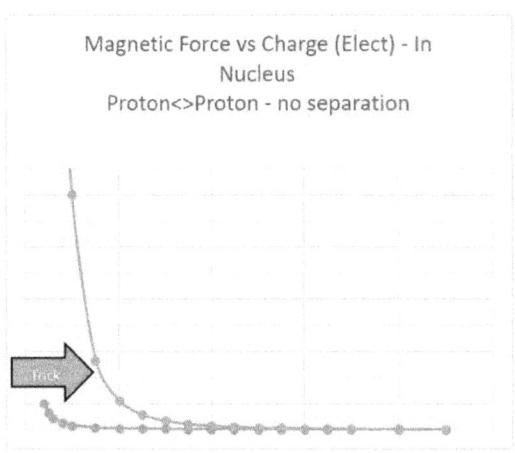

Because a magnetic field does not start decreasing until the physical chain is broken, there is proton-neutron-proton configurations that allow a nucleus to stay bonded together.

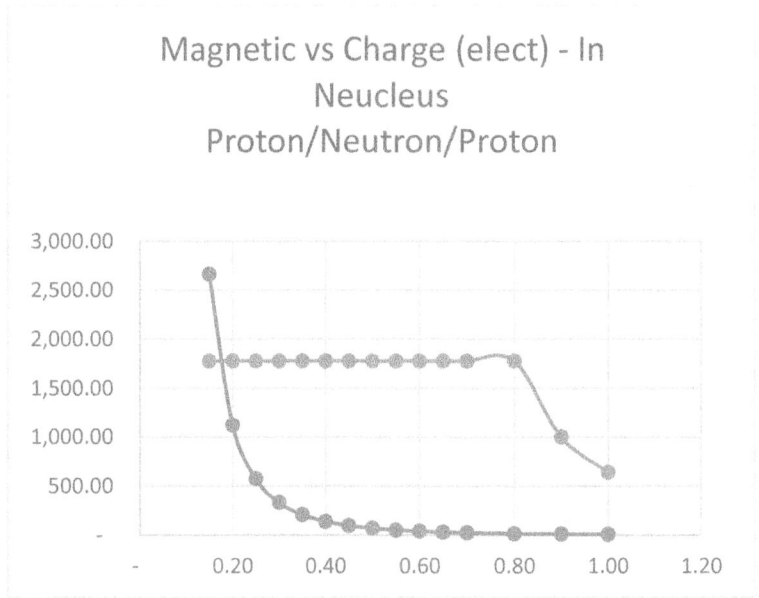

However, that long flat line only happens if there is a magnetic proton-neutron-proton chain. Now, without the neutron in the middle, the protons get very close, and proton repel at the strength of a nuclear explosion. The force gets immensely strong, even near to infinity as force calculation becomes $(\frac{1}{0^2})$.

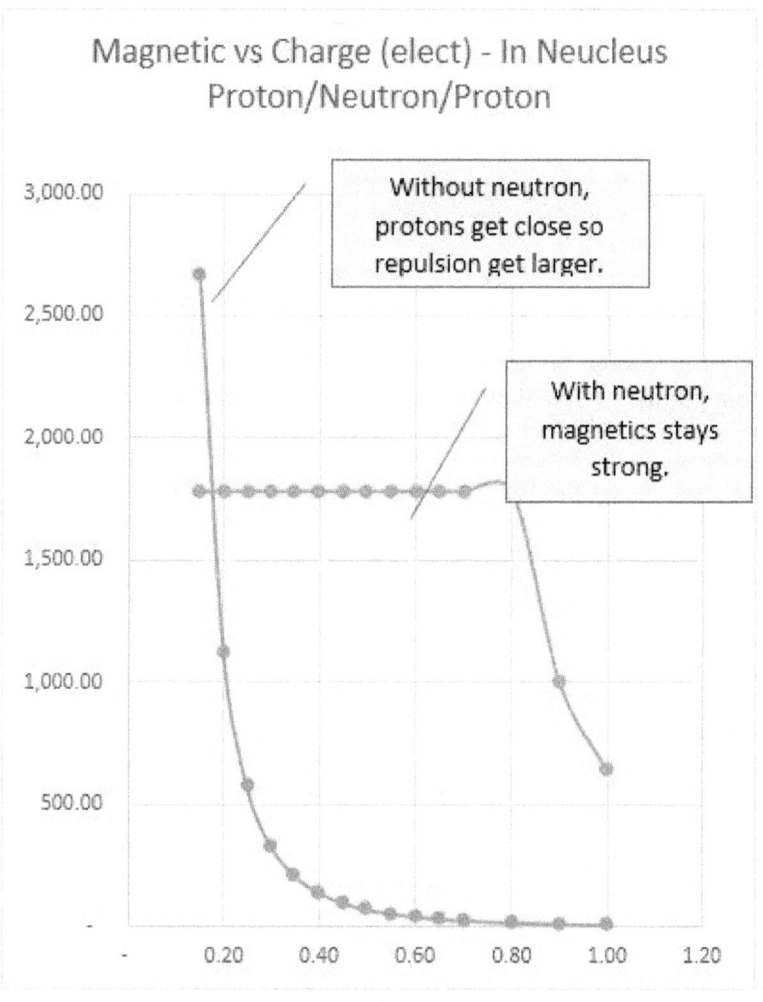

#3 Electrons shells build in geometric forms with magnetic axis repulsion creating favored positions, and not in the direct filling order of aufbau/Pauli

- Educational chemistry set patent 15245326 pending
- Resolves 006-C Carbon 109.5 angle versus 007-N Nitrogen 107.5 angle versus 008-O Oxygen 104.5 angle
- Resolves 027-Co Cobalt melting point
- Resolves 005-B Boron bonding angles at 120 degrees
- Resolves the 029-Cu Copper and other transition metal electromagnetic spectrum evidence why only 1 4s electron
- Replaces s/p/d/f with geometric m/e/c/t/v with e as intermediate in some elements. m2 = magnetic poles scrunched, e = equatorial, c6 = rest of cube with m2, and such
- Explains electrical resistance by a 3-electron equatorial shell

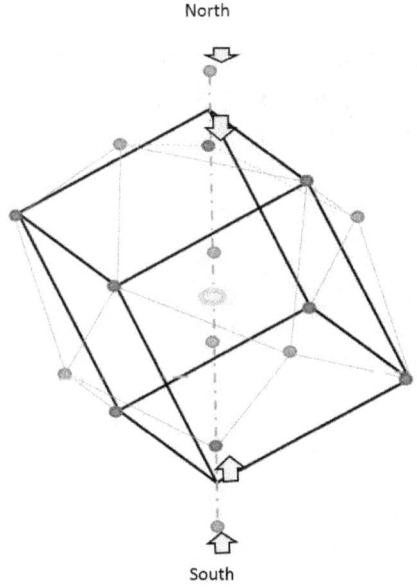

#4 Gravity is the nucleus (proton, neutron) magnetic field reflected via an almost* universal electron shell radius (R_{ES}).

- Permanently links gravity to the basic electromagnetism
- Resolves link of nucleus 'atomic mass' particles to 1/distance-squared observations of gravity
- Resolves mass loss in bonding by the bonds masking of R_{ES}.
- Replaces a number of the time-space warping calculations with physical factors that change R_{ES}, a physical distance so the calculation is knowable with physical dimensions

$$Strength = K - \left(\frac{\int_0^{Z+N} M(\#_{ZN})}{\frac{4}{3}\pi R_{ES}^3 x^3} \right)$$

$$m = \frac{\int_0^{Z+N} \left(\frac{M(\#_{ZN})}{\frac{8}{3} R_{ES}^3} \right)}{\sqrt{G}}$$

This leads to the gravity force being the charge-magnetic force for electrons in the shell particles that sits 'just a little closer' than nucleus particles. It looks like the net of two just separated is the gravity force (green):

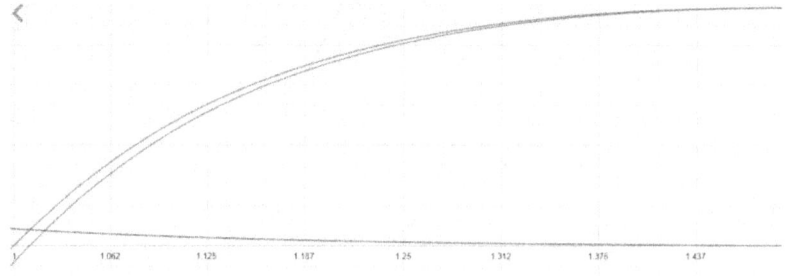

Endnotes

[i] I am using the formal mathematics term here. It has a special meaning that the force changes in understandable steps without jumps.

[ii] https://en.wikipedia.org/wiki/Neutrino

[iii] This is not to say the photon and neutrinos do not exist. However, I am working on the middle layer here. I resolve the core of chemistry using the simplest terms to be useful for certain basic calculations – like chemistry bonding, chemistry reaction priorities and speed.

[iv] Further, the postulate will identify solutions that include this operator called the 'imaginary number' which is:

$$i = \sqrt{-1}$$

Many current solutions ignore discontinuities by inserting this 'i' then applying some other factor. These factors are called 'spin', 'color', 'magnetic moment' and such although they do not have a direct relationship to either.

Therefore, we end up with a string of other factors that are discontinuous. The solutions use these complex matrix mathematics.

[v] Albert Einstein

[vi] Albert Einstein

[vii] However, that the formula for the two fundamental base forces are different ($\frac{1}{d^3}$ vs $\frac{1}{d^2}$) is not exactly true. That graph of the decrease for magnetism does not take into account the direction. The 1/distance cubed ($\frac{1}{d^3}$) is only observed in one direction, mostly because we only view magnetics north and south. It is balanced with a force of ($\frac{\sqrt{2}}{d^3}$) at the equator, and some function in between.

Yet, that minimum magnetics, which is the one, we will manipulate for the five continuous fundamental forces. We don't need the other section until we get to electromagnetic spectrums, the source of many of the Newton-reaction excess force. Just note, that ($\frac{1}{d^2}$) spherical charge versus ($\frac{1}{d^3}$) minimum magnetism going north-south does not break Newton's 3rd; the extra energy is magnetic fields pointing toward the exterior/equator (and in many cases, the photon or neutrino ejections that extra force causes).

A later chapter will introduce a 3D function for magnetism that maintains total strength offsetting balance. I will revisit the more complete function of magnetism where if you take all directions, the magnetic field is of the same total strength as the charge field. Please be patient, but understand that I am building toward that concept. That is part of the veil covering the basic relationship of charge and magnetism.

[viii] There is great complexity in calculating the electron shell distance. Even if we get the forces correct, we have adjustments:

- Angular momentum of the electrons which changes the radius distance.

- Statistical scattering which makes the force different because the distance is not a perfect circle or ellipse.

These are too complex for this stage. The calculations have a variability based upon those factors. For this stage, we look at the average electron distance as a basic constant. The roadmap for that first level of calculation comes in that later chapter.

[ix] That force gets modified for 'mass deficit' to the extent that those magnetic fields do not orient fully. Again, a later chapter to explain.

[x] http://study.com/academy/lesson/strong-force-definition-equation-examples.html

[xi] www.cyberphysics.co.uk

[xii] https://dc.edu.au/wp-content/uploads/strong-nuclear-force-distance.png

[xiii] There is a tiny variance on that charge zero in that the balancing charge elements of a neutron do have some displacement. However, that calculation is beyond the scope of this presentation. The nature of neutrons is a book unto itself.

[xiv] However, the radius (distance) of the average electron in the electron shell is linear to the Magnetics, so that is the one that Newton choose that ratio ('mass'). So, the magnetics equation that usually gets shown for this distance range – distant objects.
The radius (distance) of the average electron in the electron shell is not linear to the Charge. Therefore, calculations are overly complex, and often bypassed versus the easier linear relationship of magnetics.

[xv] A more complex validation of the electromagnetic spectrum does not show 001-H Hydrogen from 001-H Deutrium which has twice the magnetics, but not ½ the energy. The spectrum for deuterium is covered by Bohr's calculation and is only 1% different in wavelength.
Of course, the electromagnetic spectrum is based upon moving from one shell to another. That means even through you have 001-H at 1 proton, you have four (4) or six (6) electrons in the systems. You cannot have the energy release unless you have a full shell below.
That is the electromagnetic spectrum is not the Radius, but the different between different the radius of one shell versus another.

[xvi] http://www.antonine-education.co.uk/Pages/Physics_1/Particles/PP05/Particles_page_5.htm

[xvii] We engineer within four fundamental dimensions in fields. These come from real life:

Engineering Function	Everyday Description
$1/d^0$ or d^0 or 1	Location
$1/d^1$ or d^1	Speed
$1/d^2$ or d^2	Acceleration – Force
$1/d^3$ or d^3	Field Strength

These four steps are interconnected averages. The average (integral) of location changes is speed. The average (integral) of speed changes is acceleration, the average of acceleration changes is field strength.

If you know only one fact about $1/d^2$ it is that it is the average (integral) of $1/d^3$ in your direction, and $1/d^3$ is a perfectly even wave distribution process. If energy is created at a point, and goes is every (X,Y,Z) directions, the energy at any point as the energy expands is a $1/d^3$ formula. Three directions = Third in the Exponent.

However, at distant point, as in gravity, one of those direction gets seen for all the energy. The distance points gets the all the energy from one of those direction, $d/d^3 = 1/d^2$.

Therefore, whenever you see $1/d^3$ that is the total energy field in X,Y,Z. Whenerver you see $1/d^2$ that is a formula of a distant point getting that energy from one direction. At the book progresses, you will both $1/d^3$ and $1/d^3$

There is more complexity because the transformation is really

$1/d^3 > -2/d^3$ to an engineer or a mathematician, but most people only need to understand which type of the four for the basic understanding of gravity discussed in this book.

[xviii] The total calculation is complex. It involves:

- Separating the proton and electron calculations to get atom-gravity (called mass)

- Further calculating that atom-gravity for all atoms in a celestial body separately (called an integral in math) to get celestial-body-gravity

- Calculating how far out the electrons in the electron-shells actually sit to do the first calculation cleanly. These resolve certain variances in the measurement of mass already observed.

[xix] For reference, the magnetic field gets weaker at the ends and stringing around the equator (think 'bagel'). Therefore, the calculation of that requires an orientation strength.

In the larger sense, you can understand that beyond Hydrogen, the electrons are in all directions, and as such, those orientation ups and downs even out. The calculation is that amount of energy enclosed. That calculation is much easier.

Energy of N-Electrons repulsing each other = Energy of M(P+N) surrounded magnetic field.

$$N(E)\frac{kQ}{\left(\frac{r}{n}\right)^2} = N(P,N)\left(1 + \frac{r}{n}\right)^{nt}$$

Taking the separation of electrons from each other, then the volume that separation creates (a sphere) must cover the area of the magnetic field. This calculation can get down without the 'bagel' complexity.

Electrons push each other away, but protons pull them together. Protons push each other away, but electron get attracted to them. All those particles have exactly the same Charge. To each other, once you cover more than

This calculation is similar to a famous work on Bohr radius a hundred years ago.

[xx] These parts can get calculated easily.

The first shell has little electron-electron repulsion so it isolates the proton attraction well. Example of shell size balance can get seen in the Shell-1 (1p/1m) distances found in existing standards.

[xxi] There are some challenges in applying Gaussian field calculations because it requires the results to be random. That works grate for charge which is uniform, but for the magnetic elements it does meet that criteria. Magnetism is north-south oriented.

However, for our purposes here, the result is over time, and as the field rotates so the magnetic orientation can be random in that way. So, the basics of Gaussian field apply for the general requirements needed in the present solution.

[xxii] And, that total energy of the particle magnetism is the same, and Newtonian action-reaction, to the Charge of such particle.

##

ϵ_0 is called the ***permittivity constant,*** its value is

$$\epsilon_0 = 8.85 \times 10^{-12} \frac{C^2}{N \cdot m^2}$$

[xxiii] http://feynmanlectures.caltech.edu/I_01.html

[xxiv] https://en.wikipedia.org/wiki/Chemical_bond

[xxv] Insert reference to experiments.

[xxvi] http://feynmanlectures.caltech.edu/I_01.html

www.ingramcontent.com/pod-product-compliance
Lightning Source LLC
Chambersburg PA
CBHW071427180526
45170CB00001B/248